ENGINEERING:

MECHANICAL OR MORAL SCIENCE?

ENGINEERING: MECHANICAL OR MORAL SCIENCE?

by

Lewis Stretch

Becket Publications

First published in 1986 by Becket Publications,
St Thomas House, Oxford, OX1 1SJ
ISBN 0 7289 0022 X

Typeset by
SB Datagraphics Limited, Colchester.
Printed in Great Britain by
Biddles Limited, Guildford

In the cover design, the figures, based on Leonardo's analysis of the human form, are pinned on the ubiquitous micro-chip to illustrate the thesis that unbridled mechanical power tortures and kills men and women. For the horrors of the gas chambers and the atomic bombs were made feasible by modern technology. The design's relationship to the last word in the title is a reminder that the highest intellectual triumphs can easily become a crown of thorns.

'To many saints and sinners
whose shoulders built a ladder to
a vision of a renewed Earth'

Lorna Arnold

With best wishes for keeping on to the end of the road

Lewis [illegible]

Contents

Preface

Two series of items stand out from the spate of news nowadays, by their frequency of recurrence and by their dramatic impact. The first records ever more magnificent technical achievements. The second records widespread breakdowns in even the most vital, elementary social relationships. The first puts greater power at the disposal of the savagery inspired by the second.

This discrepancy between the mechanical force at the disposal of mankind and his ability to exercise responsible, humane control over it is potentially fatal. For the power has given him a dominant influence over his evolution, or dissolution. During earlier transformations — from hunters to herdsmen to farmers to craft-based civilisations — the scale of damage the human species could inflict on itself was lower. So natural forces could heal even serious injuries in time. Now mankind can manufacture disasters massive enough to destroy itself completely. It seems eminently desirable to avoid passing a critical stage beyond which such a consummation almost inevitably occurs, whether by accident or design.

'Science fiction' may continue to conjure up external threats to Earth's inhabitants, emanating from steadily more distant regions of the universe, as reconnaissance confirms the absence of hostile forces on our nearest neighbours. More accurate, careful observers recognise that the real danger stems from internal defects in man's individual and corporate personality. The most likely scenarios in which a final cataclysm can be dressed — nuclear or biological war, racial or religious conflict, climatic or epidemic disasters — spring from his lack of self-control, and the greed for 'growth regardless' that incites it.

The capacity to do so much harm flows from the application of scientific techniques to industrial and social problems. So many in desperation vent their resentment on the apparatus, on technological education, equipment and organisations. This is futile, attacking the symptoms instead of the virus that causes the disease. It is also a pity. Self-destruction is not a *necessary* consequence of applying scientific methods to human affairs, though applying them unscientifically may well be a *sufficient* condition.

Such a Luddite retreat leads to no safe haven, without a change of heart that would make it unnecessary. The material rewards for scientific progress are too tempting for universal renunciation. Yet unless it were universal, power would stay in the hands of the barbarians, and leave the time-bomb ticking away. The best, indeed the only effective protection against the imminent dangers is a more scientific approach. More not in the form of greater amounts of more sophisticated machinery, nor even knowledge, though these may follow, but rather in a more scientific utilization of experience and resources. What is needed is higher quality in the management of technological systems, not greater quantities of ingenious toys.

For the inability to control grossly powerful industrial systems is due to the unbalanced way in which the profit from previous advances has been distributed. Too high a proportion has gone into increasing output and consumption needlessly. Too little has been fed back into stimulating deeper understanding of the processes involved, and developing skill in managing them.

The degree to which specialisation has been pushed in pursuit of the former has aggravated the latter. By now the divergence of interests and attitudes between different castes makes it very difficult to cultivate sufficient breadth of comprehension for them all to be properly integrated.

The need for more serious consideration of how to maintain control of technological developments is well illustrated by trends in the central, scientific art of engineering. Too many forget that the founding charter granted to non-military engineers charged Mechanical Science with promoting 'the art of directing the Great Sources of Power in Nature *for the use and convenience of man*'. This obligation manifestly presumes the existence of some criteria that will enable the benefit to mankind of any proposal to be assessed. So it imposes a duty to ensure that any standards used for this purpose are preserved and refurbished as conscientiously as those used in the mechanical aspects of engineering, such as those of length, mass and time.

As this profession is the power-house of the technical progress being made, indifference to or miscalculation of the social elements must endanger social stability as surely as errors in design or operation endanger physical integrity.

Responsibility for monitoring the behaviour of the system rests on all who would profit from the application of science. This group now embraces almost every member of the human race. The vast, lay majority usually, rightly has a dominant voice in directing the operations of its technical services. This role cannot be safely and efficiently discharged unless they are prepared to show a modicum of interest in the characteristics of the system they want and affect to control. To gather a golden harvest they must help to maintain conditions in which the goose can lay sound rather than rotten eggs.

The duty sits even more squarely on the shoulders of engineers. For assumption of the title 'profession' implies that they have some higher motivation than their own financial rewards. They must — individually and corporately — show that they are committed to maintaining and developing high standards of professional conduct.

The author enjoyed an interesting and rewarding career that covered a wide range of tasks and institutions in engineering. It included the occupation of several advantageous 'triple points' in energy based industries. There the personal, social and technical facets converge and react so strongly that indifference to their relationship is impossible. The feedback of some experience that may help forward the development of reforms necessary for containing the hazards that any advance entails seems but due consideration for a comfortable retirement. Yet the obligation has deeper roots.

In the first place, the ability to hold such positions owed less to any inborn qualities than to the good fortune of being educated under one of the last of the great line of clerical headmasters who once inspired English education, and the brilliant team of mathematicians he led, at one of the chain of direct grant grammar schools that one flourished so abundantly and fruitfully in Lancashire. Such was the excellence of the foundations laid there that it took other very able mathematicians at Cambridge a couple of years to discover that a more practical vocation was nearer the author's mark.

In the second, these men allowed no dangerously false distinctions to be drawn between the ethical and the technical aspects of one's work. Aiming at the highest attainable standards of competence and integrity was not merely a boring necessity for passing examinations and getting a good job. It was a basic moral imperative. As strict methods had to be applied to social as to scientific problems.

How much one benefited from such influences is made very apparent by comparison with what is now being offered to one's children and grandchildren. Schools like Merchants are under threat; many have already been liquidated. Men like J.B. Lancelot and C.F. Russell no longer grace the system. Indeed the strain seems to have been lost. Quality is being sacrificed to an intemperate demand for quantity. So it seems timely for anyone who has enjoyed such great privileges to try to hand on some of the insights they provided for him before too much of value is absent-mindedly jettisoned, or deliberately destroyed.

Today there is little public recognition of the need for feasible and coherent aims in dealing with the problems of both a nation and a planet becoming rapidly over-crowded and tightly inter-connected. There is quite inadequate concern about whether the style of technologically sustained civilization that is being generally adopted *can* be controlled for the benefit of humanity. The question of whether this desirable end can be achieved without restoring their proper authority to the Christian beliefs on which modern science was founded is allowed to go by default. It seems sensible to answer the question correctly before the system becomes an uncontrollable menace.

The conviction that firm foundation on an agreed system of values is essential is the excuse for a mere mechanic leaving his traditional pastures and, like Amos, trespassing on high priestly preserves. And for adding to the notoriously unending tale of making books.

Change — like the jungle — is neutral. Survival depends on skill in discriminating between features of an environment, ancient or modern, that sustain life or destroy it. Rejecting a guide who has shown the way to the success already attained may seem a bold, radical, rational reform; but it may stem from childish petulance and folly. Throwing away all existing charts, instruments of navigation and the steering gear is hardly an intelligent prescription for making anything that can be legitimately called 'progress' through a perplexingly turbulent patch of sea and weather.

As Burke, the architect of the political consensus on which Britain's 19th century achievements were based pointed out — 'while a state without the means of some change is without the means of its conservation, to innovate is not to reform'.

THE CONTROL OF TECHNOLOGY

Chapter 1

Why worry?

Because a strange thing has happened to mankind in its search for the New Jerusalem. No doubt New Utopia, New Atlantis, New Harmony, New Frontier, or some other novel synonym would be a more fashionable title for its objective. The text will show why the original version is preferred here. Moreover, undeniable indications that it had lost its way appeared just when it seemed to have lighted on a particularly hopeful line of approach, and was reaching a stage where knowledge of how to follow it successfully was becoming widely enough diffused to open up the possibility of rapid, continuous, universal progress.

What sort of heaven are we seeking?

The human race has been engaged for a long time in its quest for The Good Life. This ideal has always, rightly included some amelioration of the raw struggle for survival that dominates lower forms of animate creation.

Intellectuals in relatively secure, well-fed situations sometimes forget the starting point of their ascent, the ground on which the foot of their ladder rests. They disparage St. Bernard's vision of a golden city blessed with milk and honey as too primitive. One must hope that the television pictures now projected right into their homes will remind them how privileged they are, how fragile the defences against destitution and barbarism are.

During most of the known history of human existence, any improvement in its lot was fitful and slow, sparsely distributed and easily lost. Life for most men and women has indeed been solitary, poor, nasty, brutish and short. Yet when Europe had recovered from the destruction of the Roman Empire by the northern invaders, and had tamed them well enough to start rebuilding civilised institutions, it created a movement that was to change the scene dramatically.

Slowly, initially within a few enlightened circles, it forged the methods of modern science. For several centuries these only affected elites, and applications of exceptional military or political importance. Then, at about the start of the nineteenth century, an outburst of enthusiasm for applying them to production processes and social systems quite generally enabled those communities open to their influence to advance at a faster, accelerating rate.

Even the most disillusioned observer of the results today should recognize that there were reasonable grounds for believing that the application of scientifically tested knowledge and techniques more consistently and widely in industrial and political spheres could free mankind from some ancient bonds, and enrich the whole commonwealth. Despite many deeply entrenched relics of less enlightened eras, Europe and its cultural partners had not only created the necessary tools. They had established an adequate area of intellectual co-operation. This is an essential pre-requisite for the system to acquire ample information, and correlate attempts to rationalize it efficiently enough to maintain its impetus.

Some of the initial hopes were undoubtedly rather naive; but there was evidence to suggest that the system could become self-correcting in time, if it were adopted honestly and consistently. Indeed, although many areas went untouched by reform, and serious relapses took place, in that century science did make a massive contribution to the welfare of humanity. Our own attempts to correct the defects in their performance and improve on their record will suffer more than the reputation of the Victorians if we fail to acknowledge their achievements in reacting against their mistakes.

How do we hope to get there?

As one common, outstanding feature of all those countries that led the Industrial Revolution was the development of strong engineering industries, every nation is now devoting a high proportion of its resources to expanding its technological potential. They are increasing the number of engineers, scientists and mechanically skilled workers turned out by their educational systems, trying to acquire more and more advanced machinery, and striving to secure and exploit the raw materials that these consume.

This worldwide growth in engineering the application of science is manifestly achieving amazing results, both physical and social. Yet it is questionable whether these even approximate to the expectations of those who advocate and support this trend, whether the environments and social conditions created by our present style of directing the Great Sources of Power in Nature are consonant with the dedication of the art to the use and convenience of mankind.

Despite a growing input of personal endeavour and output of technical progress, the twentieth century has seen many of the prospects turn very sour. Two world wars have harrowed the heartland of the order, and demonstrated that science can be applied even more readily to destroying and enslaving men than to preserving and freeing them. While it can still offer considerable financial rewards, and commands grudging respect for the military power and political leverage it yields, science no longer enjoys the unstinted admiration of those in the community concerned with the well-being of people. It shows too many signs of becoming their tyrant instead of their servant.

Yet much current criticism fails to get to the root of the problem. Some of the reactions among the young can be discounted, as typical of their natural tendency to challenge the ideas of their elders, and the disorder generated by a period of affluence. Nevertheless, the tendency to reject the discipline of science while still demanding the benefits that flow from its application to their sustenance and comfort is disquieting when carried through into adult phases.

For the systematic application of science to the methods of providing the necessities of life is now essential, if anywhere near the present population of the globe is to enjoy a tolerable standard of living. Unless sufficient weight of informed support is committed to upholding adequate standards of responsible control over its application, the way lies open for ever greater prostitution of the skills it has developed, and the growth of conditions that might well precipitate a third, finally fatal world conflagration.

The queer transformation of one of the most hopeful co-operative enterprises ever undertaken by mankind makes it the bounden duty of all who earn their living by practising or who patronise the art of engineering to examine whether, and if so how, it can be reconciled with human freedom and progress. At present too many seek refuge either in schizophrenic indifference to the consequences of their actions, or by passing the buck to society in general, claiming that it imposes on them courses of action of which they might not personally approve.

The juxtaposition of superb technical achievement with increasing social disorder and disillusionment, particularly in the more advanced nations that have profited most from it, points to the urgency of the need for careful consideration of whether conventional methods of applying science relate the means at man's disposal intelligently to the ends he hopes to attain. Indeed whether the latter are defined accurately enough to enable him to adapt his responses promptly and sensibly in the hazardous situation precipitated by the critical change in his relationship with his environment caused by the enormous growth in his mechanical power.

Problems

The main pressures that have to be reconciled to establish acceptable codes of conduct, that maximise everyone's liberty by limiting it, so that his neighbour's similar rights do not imperil his own, and promote his welfare by balancing it sensitively against that of the society that sustains him, have been identified and explored for millenia. Yet the success of the modern scientific revolution has put peculiar difficulties in the way of recognising how fundamental principles should be applied to the novel, highly developed and interdependent society that it has helped to create. These are aggravated by the rapidly changing conditions that exist at present.

Solutions to so many everyday problems have been reduced to standard formulae that have stood the test of time that the importance of the assumptions on which they are based is easily forgotten. Thus the need for sufficient common experience and ideals on which to build co-operation throughout any free community is often ignored. Mechanisms designed to prevent the system being perverted for selfish purposes where these exist simply do not work where they do not.

Yet the very developments which have reduced the number of cases that must be referred to primary authorities increases the latter's importance. A modern international standard is not as familiar to the general public as a medieval monarch's arm or foot; but its influence is much greater and more widely spread, so that it must be chosen and protected with the greatest of care.

The horizons of human knowledge are now so vast and expanding so rapidly that comprehension through adequate personal investigation of even those immediate factors that affect one's own work directly has become quite impossible. Any attainable coverage of information and experience is either too narrow, if coherent, or scattered too thinly to be completely reliable. Even in the laboratory much must be taken on trust; many 'black boxes' unproved by personal testing must be accepted as providing accurate evidence in vital sectors. When facing the more complex problems involved in applying science in society or in studying social relationships, the factual and instrumental deficiencies are orders of magnitude greater.

More powerful mechanical methods of storing and sorting information can speed its handling; but this is only an ancillary matter. The key, limiting factor is the ability of the human brain to assimilate and analyse it critically, and thus initiate humane reactions. There is neither evidence nor reason to suggest that the capacity of this bottleneck has increased proportionately to its physical resources.

Indeed, the growing reliance on mechanical aids to intellectual performance may have a similar adverse effect on man's mental facilities to that which over indulgence and reliance on mechanical transport has already had on his physique. Modern man is in grave danger of becoming mentally as well as physically flabby.

The technical difficulties arising from the sheer quantity of knowledge now possessed are intensified by some of the social changes that have developed in consequence of or parallel with it. The constant busyness of modern life, even in periods of nominal relaxation, demands so much concentration on what one is doing that there is little time for reflecting on whether one should be doing it at all. Like the Red Queen, a lot of people exhaust themselves running hard to stay where they are.

Then the growth in size of organisations has caused the transfer of many once managerial functions to remote bureaucracies, of much skilled work to the machines that high turnovers can justify. These trends have diminished the sense of human dignity and worth, so that indifference and despair often aggravate the institutional pressures that make drawing attention to critical issues or evidence difficult.

Yet no community can expect to conduct any venture with safety and profit if it does not apply whatever system it is backing with a reasonable degree of understanding and integrity throughout its successive phases. Trying to exploit a process on principles that conflict with those on which it has been developed is a sure recipe for disaster. Unless its basic characteristics and the conditions essential for it to function properly are recognised, and respected in practice by the group selecting objectives and procedures that are compatible with them, no one should be surprised if the exercise degenerates into a shambles. Technical dexterity, while a useful adjunct, is no adequate substitute for honest direction. This is a continuing obligation as long as the benefits attendant on successful operation are desired. For while, with perseverance, mistakes in reading the map or accidental deviations from the right course can eventually be corrected, those flowing from deliberate withdrawal or failure of the will are more difficult to overcome.

This duty cannot be discharged successfully unless all concerned recognise and respect the authority on which the system they are backing is founded. Without constant, strict attention to its basic postulates and demands, they can become prisoners of private prejudices, incapable of noticing some vital flaw in their approach to the particular problems confronting them at any given time. Such blind spots have often blocked technical progress, as well as misled its application.

Even when the evidence is unavoidable that, if freedom of inquiry is not to lead to horrors such as nuclear or biological war, man may have to be placed under constraints that would reduce him to a mere manipulated element in a tightly controlled social structure, many seem willing to close their eyes to it in the hope that their own privileges will last out their own lifetime. Yet unless the possibility that applied science is not the key to freedom and the easing of tensions, but a corrosive agent, is faced honestly and *scientifically*, the failure may betray alike the profession and the community it claims and ought to serve.

The informed co-operation essential for exploiting the advantages while inhibiting the dangers of applied science must be based securely on a code that guarantees both the sensitivity and the supremacy of the human conscience. Otherwise the system will so stunt the rights of man that, if he survives at all, it will be in conditions that deny even lip-service to the idea that he is a free, responsible, individual personality, by making him the slave of the mechanical and political contrivances that he has manufactured, ostensibly for his service and pleasure.

Prospects

Such a consummation of mankind's attempt to grab too quickly too much of the fruit that the general application of scientific methods can yield would be unfortunate, as it would be fatal to the science from which they spring. Genuine science is utterly dependent on free, responsible, human inquiry. It would also be oddly irrational, for modern science was founded on a common faith capable of containing the present dangers.

The outburst of the human spirit that inspired the scientific achievements that have flowed from the Renaissance did not spring fully armed upon the scene, like its mythical patroness — Athene. It evolved out of the matrix of ideas and beliefs based on the Christian religion that had been injected into and slowly permeated European society during the previous centuries. The form it took was conditioned by the axioms and attitudes on which it was founded. Before these are abandoned, scientific integrity demands that the feasibility of sustaining all the essential elements in the system by alternative means, or of dispensing with any metaphysical foundation for necessary customs and practices at all, should be rigorously demonstrated.

So this exercise considers what conditions must be observed if humanity is to retain control of the enormous forces now being released by mechanical science, instead of being crushed by them. This is done by reviewing the evolution and nature of the pivotal, scientific art of engineering.

Many signs suggest that supports that are vital to the stability of civilised society are being undermined, or removed without appreciation of their function. Yet if any components in a structure are to be effective, their role must be explicitly acknowledged, so that they can be constantly maintained and monitored, and the full implications of relying on them realised. While the value of good habits is not to be disparaged, these alone are not enough to preserve essential features in a dynamic situation. An open, caring, rational, co-operative society — the vague objective of most, particularly the young — is bound to acquire scientific knowledge, and hence mechanical power. Unless skill in retaining control over the machinery created is developed at a comparable rate, this can as readily cripple or destroy communities as it manifestly does individuals.

The anti-social nature of much engineering activity, and

many of its products and by-products, can be accounted for most simply and reasonably (scientifically) by the thesis that elements essential for the humane control of an immensely powerful movement are being discarded with frightening casualness. This demands consideration of how such queer, inhuman, irrational, unscientific behaviour can be corrected before it does irreparable damage.

No answer to all the questions reflected in the innumerable facets, variations and ramifications of the problem is attempted. Indeed, the course of the argument should generate doubts about the wisdom of adding further chapters to the long saga of man's efforts to reach unattainable, finally 'definitive' solutions. Asking the right question is the key to making progress. 'Ask a silly question and you get a silly answer' was known long before it was translated into the now more fashionable 'garbage in, garbage out'.

Yet there are ample indications that some popular attitudes and modes of behaviour are incompatible with the operation of the scientific infrastructure and the economic superstructure on which possibly survival, certainly many aspirations depend. And it is possible to point both to a way out of this dilemma, and some of the most dangerous consequences of failure to pursue it. But if the defects that are threatening mankind are to be corrected, groups that enjoy exceptional privileges, and hence responsibilities, in modern states must recognise the authority under which these are held. Only thus will the communities in and on which they exert great influence be able to command the power of mechanical science with due regard to the safety of themselves, their children and their neighbours.

Chapter 2

Why engineering?

Engineers will not need convincing that their form of activity is now so deeply engrained throughout civilisation that it provides an ideal sample for an inquiry into how to control the whole system. Yet others may fairly expect some explanation of this line of approach to a universal problem.

Other disciplines — such as theology and philosophy — have in the past claimed the greatest breadth of interest, and a fundamental status with the right to monitor the performance of their juniors in the intellectual hierarchy. Even some who might be happy to dismiss their pre-eminence might demur at the selection of engineering as the epitome of the technological field.

These might well put forward physics as the bell-wether of the scientific flock, influencing all other sectors by the spread of its measuring techniques and so a better case to study, or medicine, as having a more immediate and striking effect on people. Yet there are good reasons for choosing this animal for dissection, and justification for suggesting that examination of its entrails may be an aid to prophecy in the 20th century.

What is engineering?

Abuse of language, particularly in the 'media of communication', has obscured the true nature of engineering. Our senior institution is defined in its founding charter as 'a society for the general advancement of Mechanical Science, and more particularly for promoting the acquisition of that species of knowledge which constitutes the profession of a Civil Engineer, being the art of directing the Great Sources of Power in Nature for the use and convenience of man'. This manifestly includes the whole field of applied science when the motive is service to the community.

Finer distinctions between those engaged in different sectors are useful in organising work, contacts and the flow of information. But unless the secondary importance of internal classification is kept constantly in mind, barriers to the circulation of ideas can distort the natural direction of progress in deference to vested interests, as too much history witnesseth.

The definition clearly embraces many applied scientists who

would neither recognise nor welcome description as engineers. Some may accept their conscription reluctantly, and with expressions of surprise that they have been practising the art for many years without being aware of it. Others may reject the charge, and try to refute it. Although one understands the reasons for such attitudes, and so must have much sympathy with them, irrationalities in current English nomenclature cannot be allowed to distort analysis of the system.

Nor should the inability of many commentators to understand and use correctly the epithets scientist, engineer and mechanic be so meekly tolerated; it misleads instead of enlightening the public. Otherwise our affectations may leave us powerless before cruder and less human, but more coherent systems of thought and social organisation. For it is the combination of qualities that confuses outsiders that has enabled engineering to play a vital part in creating advanced economies.

Rate of change

Comparison of how our present style of civilisation differs from that of previous eras shows that engineering now occupies a unique, central and growing role in it. It is the emergence of this profession that is the main factor causing the rapid rate of increase of knowledge and material wealth which is straining the social order. Every aspect of it is permeated, powered, and hence deeply influenced by the art in almost every phase.

Relics such as the pottery and sculpture, the roads and buildings of former ages show that they enjoyed the services of craftsmen as skilled as any at work today. Reference to the books of ancient authors on religion, science and politics discloses thinkers as clear and profound as any of their successors. They lived and worked in settled communities that were large and well organised. These had accumulated sufficient capital to have been able to have fed an adequate proportion back into further development, as their literature and provision for leisure activities demonstrates. In theory, these advantages should have enabled them to have started that chain reaction in acquiring knowledge that modern science sustains, and by which it is sustained. *But none of them did.* In practice, the missing link was the integrating function of the engineer. This fertile hybrid provides a firm bond between two contrasting but complementary routes to understanding nature and exploiting it, the academic and the craft. It is the interaction between these that generates the enormous power of the scientific system. Rational speculation and laboratory experience is rapidly filtered through development and testing stages to improve production techniques; while industrial advances continually provide better tools that enable research to be carried ever further and faster.

Even where 'pure' scientists man the reconnaissance units, ranging widely and probing deeply in their pursuit of new phenomena and ideas, the increasing complexity of equipment often makes a high degree of engineering skill necessary. And it is the solid phalanx of engineers following hard on their heels that consolidates their gains, and transforms yesterday's academic theses into a firm base for tomorrow's further advances.

Indeed, the engineer is often using new information in an empirical manner long before it has been analysed thoroughly enough for an adequate scientific explanation of it to have evolved, the impact of new discoveries on society is negligible until the engineer has applied them to works of construction, or production on a commercial scale has extended the benefits of the basic research throughout the community.

Whether the present rate of change is good for humanity, or even sustainable for any length of time will have to be considered. Yet it exists; and where a close correlation is apparent between the flourishing of a particular element and changes in a society, examination of that feature may well expose factors that have a major bearing on the behaviour of the whole complex, and the feasibility of controlling it.

In addition to the unique peculiarities that give it a key position in modern economies, other characteristics of engineering reinforce the argument that this is the most promising field of social activity for investigation, in order to find out why the best of intentions seem likely to pave the way to the most hellish of conclusions. Some of these are consequences of the situation that now exists, some it shares with other professions, ancient or modern, some reflect particular facets of its function and history.

Professionalism

For instance, it is now appropriate to take a profession as a typical sample of society, in view of the joint pressures of democracy and technology to reduce former class distinctions. The proletarian mass was only tolerable in pre-scientific civilizations, even then only just, when primitive levels of technical development and organisation could accommodate large numbers with low standards of skill and social responsibility. Now that machines have replaced manual labour and brute force as the main source of physical effort, even the humblest worker commands great power. If this is to be utilized safely, every member of the community must attain the high degree of mastery of his tools and the integrity of conduct that should be hallmarks of a profession.

Socially as well as physically, the circus element that was tolerable at lower levels of evolution is now incompatible with sufficient cohesion and technical competence for survival. Those who command, or who expect to share in the least degree in commanding, the immense power of a modern, technologically sustained economy cannot afford to indulge in the clowning antics of serfs enjoying a brief illusion of freedom at an occasional, candle-powered festival.

Any such relics must be carefully insulated from the control of the main system. People cannot call for a higher standard of living for a growing population, for freedom to adopt anarchical patterns of behaviour that make even the maintenance of existing standards and populations impossible, and for exemption from the consequences of their folly, all in the same breath, and escape unscathed. The genie that they have released is too ruthlessly subservient to their commands.

Many may be reluctant to adopt such a radical change of outlook, and share privileges, obligations and styles of opertation formerly reserved for small, select groups. Yet this must be done if the pressures for better educational opportunities and higher standards of living are to be satisfied without leading to chaos. Such a reform will force a lot of people out of well-worn, comfortable and familiar grooves. The inconvenience this will cause is a minor sacrifice to avert the servitude or oblivion that unbridled mechanical power would inflict on mankind.

Prudential

Critical examination of the nature of any sector to which society is directing a growing proportion of its members and resources is in any case simple prudence. The general public has a responsibility for the quality as well as the quantity of whatever services it demands. Part of the consideration for the benefits expected is making sufficient effort to understand the basic postulates of the enterprise to enable the patron to discharge his function intelligently, and make co-operation between him and his technical advisers effective.

Unless the principles on which any line of development is based are recognised and respected by all those involved in its promotion, frustration is a more likely result than any profit. Like the submerged seven-eighths of an iceberg, the mainly sub-conscious ideals on which engineering is founded are not obvious, or easily related to the visible parts. Yet they must be there to carry the weight of the brilliant pinnacles that dazzle the eye; and ignoring their existence can be disastrous.

Analysing the history and nature of one of the professions that is responsible for the most distinctive and widespread current modes of activity is indeed an elementary scientific precaution. Despite much cant about being 'forward looking' if one ignores all previous experience and traditional lore, the very essence of modern science is that all its theories are anchored securely on well established — that is historic — facts. It is this solid foundation that makes it so vastly more effective than other, earlier philosophies.

This imperative is often obscured, deliberately by some purveyors of 'news', accidentally by the well-known concern of scientists for acquiring new information, and for devising more stringent experiments for testing, extending and revising ideas. Yet no explanation is ever accepted as proved until all these stages have been completed, and except on the strength of its conformity with the results of these and other well attested studies of every vintage.

The materials that are transforming our life style emanate from observation and experiment, not the touch of the philosopher's stone; the freedom of communication now available stems from the feed-back of experience into electrical and mechanical equipment developed from it, not from perpetual motion devices. Life would be uncomfortably 'unprogressive' without the benefit of such antique inventions as fire and the wheel.

For analysing many aspects of life and nature, a longer time-base and a wider range of conditions than any individual's own experience can cover is needed to provide depth of insight and perspective. So while established concepts must be constantly challenged by reason and new evidence, new hypotheses must be treated just as rigorously as existing beliefs. The novelty of any theory is quite irrelevant scientifically. Accuracy in fitting the facts and adequacy in explaining observed relationships are the only criteria for judging its validity.

If, then, the scientific professions are to continue to yield the quality of service that they advertise, critical analysis of the conditions that have enabled them to do so in the past is essential. In carrying out such an exercise, comparing different phases can often help in detecting if there were any ingredients in successful periods that were not positively identified at the time, or have later been forgotten, or not credited with their full importance. If so a society may, like Dr. Jekyll, find itself relapsing into barbarism, through ignorance of some vital element needed to prevent regression.

Too often the intoxicating effect of the latest advances has encouraged the false assumption that factors carry no weight if they have not been noticed, fully analysed and understood. Such intemperance and lack of perspective show how deficient even scientific education and organisations can be at times, in expounding and inculcating those philosophical foundations of their disciplines that alone can equip technical ability with insight and sound judgement.

Situational

A further advantage that makes engineering a convenient field for inquiry is that it has only emerged as a distinct calling in relatively modern times. This has occurred when many other groups have been acquiring professional characteristics, and pressing on its boundaries. So it has not been able to become as ossified, or establish as tightly closed a shop as the more ancient professions. The absence of such protective devices exposes the engineer and his work more readily to effective criticism, and makes it more difficult for him to disguise his operations (particularly his errors) behind a screen of jargon.

And as the most important phases of its development have taken place since the invention of printing, the history of engineering is better documented than those that have to be traced back through the mists of antiquity. So misunderstandings due to having to translate ideas from dead languages or totally different social systems are minimised. Adequate well attested evidence offers better material for analysis than any scattered over a wider and more varied field, but which is of doubtful quality in vital patches.

Moreover, as his work is concerned with growth and development, the main interests of normal people, it induces a more healthy and balanced outlook on his fellow men and their affairs than those vocations that deal mainly with people in disordered states, or engaged in anti-social activities.

The type of work on which the applied sciences are engaged adds another advantage. The extensive distribution of his products forces the engineer to maintain close contact with all sorts and conditions of people, and appreciate many widely differing patterns of behaviour. In trying to reconcile the conflicting pressures they impose on his projects he has to adopt a broader conception of the limits of his personal responsibilities than is accepted in other callings.

Paradoxically, engineering is also a useful starting point from which to approach the more complex, total situation within which it operates, just because it is primarily concerned with relatively simple mechanical systems. These expose in a very direct, unmisrepresentable way the cost of failing to maintain adequate standards of competence and integrity. In more complicated problems the identification of responsibility for results becomes more difficult. In social affairs the intervention of another person often corrects mistakes, and so prevents lessons from even being read, let alone understood and digested.

So while the potential dangers of selecting the wrong response from a wide array of possible reactions to a bewildering display of useful information buried amongst a mass of distracting noise grow rapidly, the urgency of curbing the murderous influence of ignorance and passion escapes notice. Like a watchman at his post or a look-out in the bows, the mechanic may not be the brightest or most highly respected member of the team; but he does occupy a vantage point, so warnings from that quarter cannot be safely ignored.

Educational

Again, engineering, as an art committed to the application of science, provides a bridge between the two poles of intellectual awareness (art and science in their narrower connotations) that are threatening to destroy our culture by tearing it apart. Its very existence demonstrates that these two faculties are not incompatible, that the observance of the strictest scientific standards is not an obstacle to aesthetic satisfaction, but can indeed be an aid. It should be a lively reminder that the conventional dichotomy into art and science subjects cherished by some institutions is an unfortunate historical accident, that often limits and distorts the entry into and the development of schools so branded. Correctly used the epithets distinguish methods of approach rather than separate territories.

Although engineering must always use knowledge scientifically, and is never free to relieve its feelings in bouts of pure emotion, it transcends the formal limits of pure science. Intuition and tradition, perception and imagination must be compounded with its analytical techniques if its solutions to problems are to contribute in the fullest measure to the convenience of mankind. Scientific formulae alone can no more design a machine or structure well than they can write a novel or a symphony, though disregard for the technical limitations of the medium employed is usually more immediately obvious in engineering than in some of the 'fine' arts.

Moreover the fact that conditions imposed in his specifications prevent the engineer from excluding many awkward features when proffering solutions to his assignments has a liberalising influence on his training. The complexity of the subject is increasing as fast as any other, and so forcing practitioners into steadily higher degrees of specialisation. Yet every aspect, social as well as technical, must be re-integrated into designs that satisfy all tolerably well if his plans are to work and be acceptable. Moreover, the wide fields of ignorance still abounding, and the peculiarities (if perverseness is too uncharitable) of those who will operate or enjoy his products, can never be forgotten. These obligations prevent the engineer from adopting the narrowness of outlook that mars the contribution of many specialized groups to the commonwealth, and threatens the vital coherence of society today.

As sensitivity in discriminating between spheres where mechanical laws set tight limits to feasible solutions, and those where other considerations carry greater weight, is an essential part of the education of engineers, it is surprising that the value of this useful and humane discipline for many who may have no intention of practising it is not recognised more generally.

In conclusion

Two further considerations might be drawn to the attention of any who remain doubtful about the central role claimed for engineering in a modern society.

The cynical might reflect that anything that enables an uncouth mob of mechanics to promote the welfare of mankind must be the very elixir of enlightenment, so discovering it would be of value in every walk of life.

The charitable might note that the situations investigated are common to all those professions that apply scientific (in the broadest sense) principles to the solution of their problems. They could then excuse the inevitable bias of any author, and translate the engineering terminology into that of their own vocation — for engineer substituting farmer, pastor, lawyer or physician, and converting mechanical science into agriculture, theology, jurisprudence or medicine.

Similarly the charge that the account relies too much on British experience must be admitted, and rebutted by the observation that as the United Kingdom played a leading part in launching the profession, it has naturally hit some of the problems of maturity earlier than other nations.

When adherents of other disciplines notice that many of the scenarios examined correspond to ones well known in their own fields, they could begin to suspect that all the substance of the argument had been filched from other archives, and simply dressed up in a boiler-suit. Yet such competition for priority would be unprofitable.

The true explanation is that there is an underlying unity in all human experience and patterns of thought. This has been obscured by the over-specialisation which is one of the less happy relics of Europe's nineteenth century wealth, now deeply entrenched in many circles. Different facets of a problem are not separable in *real* life. They are not independent phenomena, but features whose interaction creates the experience. So their relationships must be included in any full account that will enable the significance of each to be understood. Units in turn are elements of one inter-connected universe, and so related to and affected by the remainder.

So it is not surprising that methods of analysis that have worked well in one sector are often found to be equally useful in others, when they have been translated into the local dialect, and modified to fit a different set of coordinates. Today it is expedient to set them out in technological terms, and test their relevance in industrial circumstances, as these are the main centres of interest and activity for the majority of those living in advanced communities. Such an exercise is also essential if there is to be any hope of repatriating some of the extremely powerful techniques that have been developed in those fields to help in dealing with the control problems of a growing population that must live together if it is to live at all.

His selection as the median characteristic of modern societies is not a claim for privileged status for the engineer. It simply emphasises that the occupation of a key position in the social system at any stage in its evolution by any profession puts a high responsibility on it, not only for the conduct of its own operations, but also for explaining to allies and the public in general the inherent possibilities and limitations of the art. The present nonsensical mixture of over indulgence in some spheres with the neglect of their advice in others shows that engineers have failed to make the nature and capabilities of their calling clear to their fellow citizens.

The applied sciences are neither a set of clever tricks, regularly updated, that any smart, ambitious type can master, nor a compendium of infallible formulae that any properly briefed high priest can dispense. They are empirical, evolutionary disciplines that only yield genuine benefits to those who follow them patiently and soberly.

Engineers cannot apply mechanical science for the use and convenience of their fellows if the latter insist on destroying the instruments and environment necessary for defining their objectives accurately, and enabling the profession to work effectively towards them. Even in democracies, 'Caesar' can only receive honest tribute if he refrains from debasing his own currency. So the boffins in his mint must point out the consequences of trying to defy the laws of nature, social as well as physical, even at the cost of some unpopularity, and the risk of being accused of arrogance. A wet engineer running naked through the streets shouting 'Eureka' may not be any respectable monarch's idea of gentlemanly behaviour; but it did save him from being taken for a ride.

THE NATURE OF THE ART

Chapter 3

Evolution

The basic, original, primary function of engineering is to provide the tools required for feeding mankind, and protecting him from rivals and the elements. Extending the expertise it has acquired in order to satisfy more sophisticated demands is a secondary, later development. In the early stages its products were so elementary and so commonplace that it took a long time for the art involved in their manufacture to be recognized as a distinct vocation; every man had to acquire it if he were to survive.

On the other hand, the ancient professions deal mainly with less common aspects of living and communal relationships. So they had to create much earlier the specialised schools and apparatus that the public now expects to find as characteristics of professional groups. Indeed one of the difficulties that still afflicts the definition and organisation of engineering is that it never can, nor should ever try to disentangle itself as completely from the day to day work of ordinary people as the older professions.

So when it did become recognized as a distinct vocation during the Industrial Revolution, it inherited a vast fund of knowledge and traditional practices from its predecessors. Much of this was widely diffused and commonly practised. But two more specialised methods of mastering natural resources contributed the experience they had been accumulating and the ideas they had been testing over centuries.

Craftsmanship

The senior in time was the craft tradition. This stretches back in unbroken line to the simplest mechanical skills that first lifted man above the animal world. Its present, apparent submergence by its younger, now more massive, academic co-tributary is quite illusory. Progress is still dependent on the availability of a wide distribution of practical skill throughout any community. This is essential for translating theoretical designs into machinery, and then operating and maintaining them efficiently. The spread of general education, and the development of improved methods of storing information, disseminating it, and training both specialists and the public, must not be allowed to obscure the vital part that craftsmanship and manual dexterity still play. The general diffusion of a considerable degree of mechanical competence is indispensable in any society whose way of life is sustained by engineering.

This requirement applies not only to the workers on the shop floor, and the laymen who enjoy and operate its products. Practical experience is necessary at all levels of management up to the very highest. It provides insight that is an essential complement to analytical methods; for seeing a giraffe can save wasting a lot of time in proving that the animal cannot exist. These technical reasons for respecting manual modes of learning and doing are reinforced by many historic warnings of the social instability or obsolescence that attacks societies that allow rigid caste systems to paralyse their flexibility of response to changing circumstances.

Once a significant amount of knowledge had been acquired, institutional arrangements for preserving it and handing it on had to be developed. The style of these varied widely, and hence their strength and influence. Craftsmen being but human, one factor promoting these was the selfish itch to restrict access to valuable skills, so that they could be exploited for the maximum benefit to current possessors. Just enough was allowed to pass down to the rising generation to provide for one's old age.

Yet from the earliest known instances, other factors were also admitted. The importance to the community of the services rendered by the crafts ensured that an element of social responsibility was introduced from the start. The fascination of the arts themselves also inspired their own mystiques, and preservation groups. The divergent pressures from different sectors of the community were sometimes reconciled by externally applied sanctions, sometimes by prophetic insight, both were usually necessary.

In those phases of growth when the skilled trades played a major part in freeing mankind from physical and social chains, it is pertinent to note how deeply their structures were permeated by ideas drawn from contemporary religious ferment. The craftsman realised that his own freedom and status depended on his acceptance of a concept of service that imposed moral obligations. Only within such a framework could the practices needed to safeguard his own speciality be justified, designed with sensitivity, implemented effectively within the craft, and tolerated by those outside.

The way in which this obligation was acknowledged varied. In classical and medieval times it took the form of direct affiliation of the guilds to patron gods or saints. During the Industrial Revolution it worked through the less formal influence of the 'nonconformist conscience'. Even today, when most of the ideals that gave them significance have been discarded, traces of these origins remain engrained in union conventions. The most intriguing one is the custom of calling one another 'Brother', long after any feeling of having one and the same Father has been rejected.

Scientific power

The second stream was the scientific one. This stimulated the disciplined imagination needed to deepen understanding and widen the boundaries of the art. Although some strands go back many centuries, the present form stems from the intellectual reforms that were initiated at the Renaissance. These have been developed since then by those communities that have been open to their influence.

This re-birth of learning has often been misrepresented as a sudden, dramatic break in the continuity of human evolution, triggered by the reflux of classical texts into Europe. Yet these had always been available, had anyone been anxious to obtain and consult them. The factors that precipitated the rapid change then experienced were the recovery of sufficient defensive skill and social cohesion to enable communities to protect themselves from marauding bands, the growth of the economy to a level that allowed patrons and scholars to divert a greater proportion of their interest and resources from subsistence and defence to more civilised pursuits, and the technical advances that made these developments feasible.

Society certainly went through a critical change of state at this period. Yet this did not occur in the catastrophic way that sometimes makes it difficult to recognise the continuity in natural processes. It was achieved — as all beneficial, well organised approaches to criticality are — by a slow bringing together of all the essential ingredients. So neither the significance of the change nor the exact moment of transformation was detectable at the time.

For many centuries men had had little opportunity for intensive study, reflection and writing. Nor had they enjoyed any cheap methods of reproducing texts, to speed the exchange of ideas and critical appraisal, and to increase any work's chances of survival in troubled times. So the first printing presses had perforce to rely mainly on classical authors for a ready supply of worthwhile material. Even so, the most revolutionary impulse did not come from anything salvaged from Byzantium. It sprang from the translation of the 'Vulgar' Bible into the language of the people, and its general distribution.

Its teaching, which had been slowly penetrating and taming the fertile but savage enthusiasm of the European tribes in second-hand forms, was the most powerful influence in the movement. Its demand for both personal freedom and the acceptance of the conditions that enable its exercise to further the welfare of the community instead of degenerating into anarchy provided the charter alike for the scientific system and the concern for social improvement that have gone hand in hand and flourished together ever since.

It was from the preceding 'Age of Faith' that the men who launched modern science inherited the belief in the essential unity of the world, and of all human experience. This enabled them to dispose of the inadequate, incoherent theories that had been handed down to them. They then used these basically religious axioms to start the process of discovering the order underlying the apparently chaotic diversity of nature.

The same authority justified both their empirical methods and their assurance that their efforts to elucidate the laws of nature were not merely an amusing pastime, but a sacrifice of their talents as acceptable to their Creator as the customary exercises of their clerical brethren. In maintaining this wholesome position they followed their Master more faithfully than those who abandoned to the Devil the world he loved so much and came to save; so they inherited the right to correct the errors of the priestly caste.

The scientific revolution was part of the spiritual revival that transformed European society, not an independent affair that happened to occur at the same time. Those who cut a way so successfully through a tangle of deeply rooted obstructions to lay firm foundations for the new system securely could afford no ivory tower illusions. They saw themselves, their vocation, their society and environment, and the universe they set out to explore in a unified light. In their context, this inevitably meant 'Christian'.

Union is strength

Although the two streams had always reacted on each other to some extent, fruitfully on some occasions, marginally on others, snobbery had usually kept them apart. So these two complementary ways of getting to understand the natural world, the manual and the intellectual, had to be welded firmly together in the Industrial Revolution to form the scientific art of engineering.

This development was also closely associated with a religious revival. Reference to the lives of the founding fathers shows how widespread was the sense of moral obligation in everyday business inspired by the Methodist movement. The successful introduction of scientific methods into industrial and social spheres depended on Christian attitudes and ideas having been widely adopted throughout society, as Marx's bitter taunt confirms, however distorted many of the manifestations rapidly became at the profits extractable from the process corrupted its beneficiaries.

It was from these roots that the commitment of the profession to the use and convenience of man grew. Too many have forgotten, or have allowed themselves to be conned into denying, the source of the ideals that awakened society to the challenge of trying to improve the quality of life by the application of scientific methods. These also provided the impetus for establishing the more rationally ordered patterns of behaviour necessary if this were to be achieved. So one should not be surprised that the system behaves so perversely at the present time.

A free and creative atmosphere for developing new ideas and techniques, and a co-operative outlook in society for facilitating their trial and assessment, were essential conditions for carrying through such a reform. Unless the fundamental beliefs of the community had been compatible with, indeed had stimulated such attitides, it would have been unable to sustain the degree of concentrated effort needed over a long period for significant gains to be made, and consolidated.

It might have adopted them as a passing fashion for a time, and abandoned them just as easily, without leaving any permanent impression. It is interesting to stumble across many intriguing 'might-have-beens' in history, when great intellects or opportunities appeared in unfavourable circumstances, so that interest and grip were lost when setbacks occurred.

Deep faith in the ultimate coherence of the universe, in the ability of the human mind to discover the laws that govern it, and in the value of trying to do so and to apply them in the service of the community are essential. Without these successive generations lack the inspiration and the network of relationships needed to resolve continuously, year after year, century after century, the anomalies that break each partial solution, each closer approximation or more comprehensive formulation in turn. No single sector can forge ahead very far by itself. Particularly in the applied sciences, interaction of branch with branch is necessary for providing each with new ideas, techniques, materials and instruments.

Later consideration must be given to whether other starting points and stimuli exist, or might be conceived, that are capable of leading to the same conclusion. The inescapable fact is that the actions and precepts of those who established the profession of engineering on sound lines show that they adopted the practices as the correct way of discharging their binary Christian duties. They offered the sacrifice of their personal integrity and mental strife to God, and the service of their skill in easing the burden of toil, misery and disease to their neighbours.

The framework that gave the modern industrial system its basic ideas and good habits, and encouraged sufficient constituent parts to work together, was the Christian faith; however much it was abused by some of its nominal adherents, and ignored by others. It provided the discipline that enabled science to advance like an army, instead of dissipating its efforts like a mob.

Forgotten link

It may seem strange that such a vital connection should be so commonly denied or ignored. Yet it is very easy to miss the most critical factor in any problem, or solution. This is often an assumption made without recognition, let alone acknowledgment, before the investigation begins. Just because the society in which modern science evolved was conditioned by Christian modes of thought and action, many were quite unconscious of the fact, or of its influence on their outlook. Most assumed that their ways of thinking and behaviour were axiomatic, so they never considered the possibility of others denying them, or trying to apply customs derived from them in ways that conflicted with the primary code on which they were founded. Others probably thought that it was so obvious that all their activity depended on their beliefs that it was unnecessary to reiterate the fact on every occasion, that their regular formal witness was enough.

Consequently subsequent generations have often failed to notice the direct connection between the scientific system that Europe created and the fact that it had been reciting the Creeds, hearing and watching dramatic presentations of the Scriptures, and at least formally assenting to these statements of Christian belief for so long. The language and pattern of ideas they encapsulated had been deeply assimilated. They dominated both its thinking and its collective sub-conscious. The general understanding of their implications may have been primitive and shallow. The submission to their authority was fitful, particularly when it conflicted with individual or tribal lusts and prejudices. Yet these provided the network of postulates and relationships on which its intellectual and social reforms were founded.

In fact some did acknowledge the importance of this aspect, and record the source of their vocation. If concern for correcting the defects of over-specialisation is to give the younger generation a more liberal education, that is one meet for free or freeing men, it must enable them to gain some insight into the personalities of the giants on whose shoulders they stand, as they try to catch a glimpse of a new scientific heaven on earth.

This must be done by studying their work, as much as possible at first hand rather than through someone else's interpretation; the anonymous stone, glass, wooden and metal testimony of the medieval craftsmen, the brilliant projections of the pioneers of enlightenment from Leonardo through Bacon to Newton, and the record in artefact and on paper of how men such as Smeaton, Telford and Faraday wove all the strands together to launch engineering on its meteoric course.

They will then discover that for one Laplace omitting reference to God because he thought he did not need the concept, there were many who realised its fundamental relationship to the way they pursued their quest. It was also responsible for the stability of the community on which they depended for their sustenance, and the maintenance of the environment necessary for their work.

They may then go on to understand why these men and the societies that nurtured them were able to forge such effective tools for increasing knowledge, and appreciate the necessity for some foundation of belief on which to build any coherent system of thought and behaviour. There is no excuse for neglecting this aspect of science's credibility and effectiveness today. Bitter experience abounds of the cost of assuming that men and nations brought up under totally different cultural traditions will 'play cricket' and not do what 'is not done'. After all, cricket is not what it was.

Chapter 4

Present state

There are two starting points for an examination of any social group, in order to determine its essential characteristics. One is to identify the peculiarities of its members, in the belief that it is these personal kinks that drive them to behave in a certain manner, to meet and work together, and eventually to become recognised as a distinct calling. The other is to note the type of activity in which they indulge, in the belief that it is the demand for this form of service that attracts those with some inclination towards it, and so develops in them the attributes and among them the structures necessary for providing it. The evolutionary nature of animate creation drains such 'chicken and egg' debates of much significance, particularly when discussing sub-groups of a species that can modify its environment, and hence the forces guiding further adaption, as deliberately and drastically as man. Starting from the discipline, its specification and mode of operation is more convenient here.

Character, in theory

The definition used here of the profession of engineering is that set out in the original charter of the first non-military institution, the Civil Engineers, with some minimal recent glossing. As with many other forms of literary composition, such as liturgies, epic poems and plays, those doing a good job when the demand first emerges and the field is green enjoy considerable advantages. Omitting lists of examples as dated, and adding clause numbers for later reference, this states that the Institution is to be:

1. a Society for the General advancement of Mechanical Science,
2. and more particularly for promoting the acquisition of that species of knowledge which constitutes the profession of a Civil Engineer,
3. being the art of directing the Great Sources of Power in Nature for the use and convenience of man,
4. as a means of production and of traffic in states both for external and internal trade,

5. as applied in the construction of...for internal intercourse and exchange,
6. and in the construction of...and in the art of navigation by artificial power for the purposes of commerce,
7. and in the construction and adaption of machinery,
8. and in the drainage of cities and towns.

Comment on three facets of this masterpiece, that have required modification with changes in circumstances, is needed before high-lighting the abiding features that are essential, if a group so defined is to remain a coherent profession, and socially tolerable:

a) Mechanical is used in (1) to distinguish between an interest in devices rather than in processes or organisms, between applied physics and applied chemistry or biology. However, the development of all the sciences has reduced the significance of ancient, traditional boundaries, indeed even those between scientific vocations and many that would not have been classified as such formerly.

b) Drainage in (8) was simply the sphere in which the responsibility of engineers for providing a healthy environment was first recognised; though the exclusion of rural, agricultural projects reveals an insular blind spot, in view of the history of the subject. The development of industry has necessitated the extension of this clause to cover the 'care for the environment and the interests of public health and safety'.

c) Similarly the absence of any reference to economy reflects the limited vision, in its early days, of the extent of the impact that engineering would eventually have on the surface of the Earth. This has been revised to include 'the best use of scarce resources'.

Yet these modifications do not affect the importance of the key factors in the definition. These are:

Profession

Engineering is a profession (2), not a trade or mere occupation. This term is now frequently used with its one essential element missing. Professing to follow a calling while denying the existence of anyone to call is as absurd as claiming to be 'Brothers' while denying the existence of a common parent.

Perhaps inevitably in a nation devoted to sport, many professions have followed the bad example of the sporting 'fraternity', which has in consequence ceased to be sporting. They equate professional with paid, and ignore the need for a sense of vocation. So they concentrate on formal qualifications and restrictions on entry in trying to get the status and rewards that should be the consideration for the quality of service rendered by a dedicated group of people.

Engineers should use language, like all their tools, more precisely.

Service

It is a service profession (3). Its direction of the Great Sources of Power in Nature is only genuine engineering when directed to the use and convenience of man. The most brilliant technical achievements are NOT engineering unless they satisfy this clause. The art cannot be pursued either for its own sake, or for the glorification of the artist or his patron.

Engineers often find this criterion difficult to accept. Many feel that the conditions under which they serve are unduly onerous, envy the rewards extracted by other groups, and are tempted to obtain comparable ones themselves by antisocial tactics. There is much justification for their complaints; but the remedy must be to remind society that it gets the quality of service for which is pays, not to betray the nature of their calling.

Power in the service of mankind, the whole species, neither unbridled power nor power in the service of individuals or sects, is the only legitimate object of the engineer's efforts, his only title to respect, or even tolerance.

Science

It is a scientific profession (1). This both emancipates the engineer, and imposes constraints.

He can never be satisfied with a purely conventional way of tackling problems, however effective it may have proved in the past. Nor can he adopt the arbitrary, emotional methods of making choices popular in other quarters. He must have a rational basis for his decisions, or go on searching for one.

This constant probing is just as vital in exploring the useful limits of established practices as it is in testing the validity of novel theories. Under such a discipline he is free to follow any theory that fits the facts well beyond the bounds of established custom. So the scope and power of the art can be extended, and its pool of ideas and techniques refined and reinforced by analysis, and then projection.

This demands careful distinction between propositions that are truly absurd, and those that so disturb existing patterns of thought that the effort needed to comprehend and assimilate them is distasteful, hence unreasonable, disloyal, blasphemous, or whatever abusive epithet seems most appropriate in any given context, or comes most readily to mind.

The imagination and intellectual curiosity that inspires the whole scientific field is part of the engineer's heritage. Experience is a guide to warn and a friend to help, not a ball and chain to limit his freedom.

Art

It is an art (3). This also is a double-edged commitment, sometimes enlarging, sometimes restricting the range of options open to the engineer.

As it occurs in conjunction with the advancement of science, it confirms the contention that art and science are not basically opposed, but complementary. What it emphasises is that most engineering problems involve the evaluation of some factors that are not amenable to scientific analysis. These require viewing from a different angle, and assessment by a different calculus.

Moreover, they can sometimes be solved by well proven customary techniques that have not yet been fully, scientifically explained. These are none the less available, and to be used if they offer the best solution.

Comprehensiveness

It is a comprehensive art (3, 4). Its dedication to the convenience of mankind leaves none but the most determinedly isolated anchorite outside its field of concern. The specification of the objectives of its operations also means that no factor which can influence the achievement of these in the very complex environment it has helped to create can be ignored.

The 'pure' scientist can, frequently does, often has to exclude awkward elements to enable his theses to be defended rigorously. To this end he designs his experiments to exclude as far as possible all but the specific aspects in which he is interested.

The engineer cannot disregard anything that may prevent his designs from working properly, from satisfying his clients, or from respecting the rights of members of the public on whom his work may impinge.

Constructiveness

It is a constructive art (5, 6, 7). Its primary concern is with the provision of the physical works, the machinery, processes and systems used in human intercourse and commerce. Its interest in theories is secondary, as a means of attaining its objectives and improving its performance.

The acid test of good engineering is how fully it satisfies the aspirations of those who commision its works, rather than how fully all facets of the reasons why it does so are understood. All these diverse elements must be integrated. So the job specification of engineers must include vocation and utility, science and art. This is quite contrary to some fashionable trends, particularly to the strong bureaucratic pressures towards mechanical classification and feudal regimentation that abound. Yet finer resolution is well beyond the power of the experience and insight available at this stage of our evolution. So to try to segregate into corrals separated by stock-proof fences groups that must overlap and interact if they are to retain any vitality only hamstrings professional development.

New words, such as technology for engineering, are introduced that are redundant, still inadequate, and clutter communications with echoes from battles long ago. Meanwhile the coherence of both professions and society is threatened by conflicts of interest between more and more sects, founded to embody rapidly multiplying distinctions that make poor old Polonius' categorisation of the wandering players sound terse and rational.

It is better to retain a simple description of the species, and hope to identify members by observation of their behaviour. Honest or contrived errors may then occasionally allow some characters not properly qualified to assume a title they do not deserve. Yet the alternative is to arm oneself so heavily with badges of rank and regiment that the constant reference to the code book needed to distinguish friend from foe distracts attention from more serious matters. Remember Goliath!

Behaviour, in practice

Mechanical science claims to be a true component in the entire, continuous spectrum of scientific disciplines. So it is necessary to consider whether any of the additional obligations to which it subscribes impose constraints that other scientists might challenge as, at best, irrelevant or, at worst, incompatible with their idea of the purity of science. In fact the distinction between pure and applied sciences is now of little significance and doubtful validity. Yet the shadow of its passing still casts a baleful influence in some quarters, so it cannot be ignored.

Two main features distinguish modern science from other, earlier, less effective systems. These are its insistence:

1. That the only legitimate method of proving theses is by experimental verification; that theories must fit facts, not *vice versa.*
2. On the open, as rapid as possible, publication of data and deductions; so that the judgement of one's peers on the accuracy of the information and the soundness of the argument can be the authority for acceptance or rejection, not their compatibility with some existing, infallible person or caste, code or procedure.

Publication

The demand for uncensored publication is not just a transfer of the privilege of promulgating their ideas from one high-priestly elite to another more widely distributed and fashionable one. Nor is it a device to flatter the ego of the scientist, in the hope it will make him perform better. It is an essential part of the process, and so a duty imposed on anyone claiming to observe scientific standards. No scientist has completed his investigation and discharged his personal obligations fully until, as far as he is able, the work has been exposed to the world in general, as well as to his immediate colleagues.

Only thus can observations and arguments be properly assessed. Checking must be done not only by others engaged in the same or closely related fields, for they may suffer from similar biases and blind spots. Theories must be viewed from a multitude of different angles, any one of which may expose unjustified assumptions or errors in the analysis. Similarly his techniques and measurements must be checked against and collated with others from many different stations. Any remote one of these may hold a unique, vital clue, that enables accuracy and adequacy to be thoroughly tested.

Moreover, the demand for experimental verification is not limited to fitting known situations. It requires a continual search for better, more penetrating methods of demonstrating that the theory proffered fits a wider range of phenomena more closely than any of its rivals.

Consequently these methods can never lead to the absolute, final, incontrovertible certainty claimed for other systems. More refined instruments or astute observers may at any time detect new evidence or old flaws that demand re-assessment of existing ideas. However, it is only fair to note that some simple phenomena and relationships have been so extensively examined that the chances of having to abandon existing formulae that rationalise them completely, as opposed to having to modify them so that they will fit into more comprehensive ones as subsets, are so small that they can be ignored for most practical purposes. Yet such a system can only work as long as intellectual ability is coupled with the personal humility needed to recognise human limitations. Arrogance is incompatible with true science, infallibility unknown therein.

The commitment of the art to the well-being of mankind leads engineers to find most of their employment in industrial and service organisations, rather than in academic institutions. There the commercial or security value of the information and inventions they produce usually impose some delay and restraint on publication. Yet this is mainly a question of degree and time-scale.

Once a product is marketed, it is impossible to keep the reasons for its success secret. Even in defence sectors, the impossibility of maintaining completely watertight security, international pressures such as the prestige value of high scientific achievement, or even the necessity of obtaining sufficient exchange to keep one's own establishments up to the mark and prevent in-breeding, ensure that developments are revealed eventually. The delays involved can be costly to the public, unfair to some of the individuals concerned, and may cause a larger discrepancy between the ideals and practice of the profession than do similar defects in other fields of scientific activity. Yet it is doubtful if these handicaps are onerous enough to prevent most engineers from conforming — very roughly — to scientific standards.

Verification

There is, however, an inevitable, fundamental difference between the approach of engineering, indeed of all the fullscale, applied sciences, and that of laboratory science to the task of experimental verification. This stems partly from the nature, partly from the object of the exercise. The size and complexity of engineering developments makes it seldom technically or economically feasible to conduct anything like the classical critical experiment to check the theories on which they are based. In the main, only isolated aspects, elements, or sub-assemblies are amenable to testing by strict, reproduceable methods.

Yet a proper test must include all factors; so it can only be arranged on a real, full-scale project. These can seldom be isolated sufficiently from random influences to satisfy the demands of an experimental programme. Nor can they usually be repeated to check results. They alter circumstances, for better or worse, but for certain, and for ever.

To overcome some of these limitations, the applied sciences have developed methods of analogue and statistical analysis. These have been so successful that other disciplines have adopted them for dealing with many intractable problems, sometimes with considerable success. Unfortunately this has tended to generate a false belief in their infallibility, instead of stimulating the realisation of the tentative nature and inherent limits of all scientific propositions.

Moreover, when any experimental programme passes beyond the laboratory and test-bed stages to having a direct impact on the lay public and the outside world, the analysis of the results — and hence any judgement on their significance — cannot be separated from the set of values used to assess the profits and losses it entails. Even if assent is given to the belief that all knowledge is of some benefit to mankind, it does not follow that this outweighs the cost of gaining it, or that it all has applications that are beneficial. The main return may well be the knowledge that it does not and has not.

Analysing these deep and complex issues indicates the need for some higher calculus comprehensive enough to include all aspects and factors involved, and respected by all the parties to them. Unless, until it is developed, scientific judgement of many issues is impossible.

Chapter 5

Working assumptions

Despite some differences in its mode of operation, engineering shares the working assumptions on which modern science is based. So the strength or weakness of these must be taken into account in assessing whether any infrastructure is needed if a technologically sustained system is to be viable in the long run.

For modern science is not simply a convenient habit or intellectual fashion. It is a discipline, that rejects many popular methods of explaining natural phenomena and relationships. So if it is to retain its coherence it must maintain the standards it sets throughout its operations, external as well as internal.

As one cannot fool all the people all of the time, this measure of professional integrity is also essential if it is to command general respect. The obligation clearly entails ensuring that its own postulates and practices are defensible under the tests it applies to other material. So they must be demonstrably self-consistent and compatible with the facts.

Only if every kind of human experience can be elucidated without calling on any preconceived notions whatsoever is the scientist entitled to claim to be able to dispense with metaphysical support for his doctrines and behaviour. No field of enquiry can be arbitrarily excluded. If he overvalues his own prejudices, spheres of interest, and those effects examinable by the techniques in which he is skilled, he is being as irrational and dogmatic as those clerics of a similar disposition who so long, so disastrously, and so futilely opposed the incursion of science into areas that they had traditionally dominated. Then the weapons he forged for cutting their pretensions down to size will be just as effective against himself.

Prudence and probity alike indicate the need to look critically at the practices and assumptions of scientists. These are sometimes inherited from the traditions in which they are nurtured, sometimes adopted in searching for a careers to match their talents. They often grow on those opting for the disciplines quite casually, almost sub-consciously, without serious consideration of their validity.

Even if the problem crosses their minds, it is frequently dealt with by the simple, empirical reaction that they have worked well so far. 'Why?' and 'How much further?' get relegated to the bottom of the pending tray.

Yet when many well established traditions are breaking down, and the whole pattern of employment in the developed nations is changing, there is urgent need to face these questions. Even if habits have been soundly chosen in the first place, careful attention is needed if they are to be applied effectively enough to prevent the system based on them degenerating, or getting out of control in different circumstances.

Vision of the universe

Natural science assumes that the whole universe is governed by a rational, consistent and immutable set of laws. This is an amazingly arrogant and absurd assumption, unless it happens to be true.

Unfortunately the indifference of most teachers to the history of science has obscured how frequently established views of what these laws are have had to be changed to protect this central dogma. Instead of making it clear that this is an evolving process, the latest pedantry is stuffed down immature gullets.

So far from being self-evident or conclusively demonstrable, there never has been a time when much of the evidence available would not deny the belief. Every theory suffers from anomalies and inadequacies, that multiply as new instruments probing novel situations yield fresh information. As fast as one set of conflicting observations are reconciled, the further advances that the achievement makes possible bring new facts to light that do not fit into the predicted pattern.

Yet despite all appearances that nature sometimes behaves in an irrational and wayward fashion, and that the specialised branches of her study are running on divergent lines, scientists insist that this is an illusion due to limited knowledge and understanding. They maintain that every detail will eventually fit into one coherent picture, even while the number that do not grows, and ability to comprehend the whole system in order to confirm the belief becomes steadily more impossible.

The practical successes of science are indeed triumphs of faith over masses of unruly facts.

In addition, it is assumed that the human mind can comprehend these laws, that the patterns observed are not simply figments of man's imagination. The regularity apprehended might be a mental blockage inherited from common ancestral stock, or even a by-product of common building material. The wide variety in human characteristics, across both time and geography, denies the alternative explanation. Yet this still leaves the assumption unsupported by rigorous scientific proof.

The contention that these initial postulates are justified by the way in which separate pieces and previous anomalies have been steadily fitted into more comprehensive explanations that attract ever wider agreement is a natural, intuitive reaction, not a strictly deducible or demonstrable proposition. Assent to the soundness of trusting such deep-seated human instincts backed by such a pattern of evidence must admit, if the scientist is honest and consistent, equal validity for some of the basic assumptions of other disciplines. These he is often loath to accept.

Nor does abandoning strict logic and evidence for the goulash of statistics help the scientist to shake off his supernatural shadow. Once the overwhelming weight of probability argument is allowed in, it must be admitted without constraint. So it forces him to recognise moral factors and laws charted by the same process.

Methods

The present continuously successful phase of science, in which new theories tend to subsume older formulations rather than overthrow them completely, has stemmed from the adoption of inductive methods instead of deductive ones as the main scientific tools. It is this change in technique that has enabled the modern system to forge ahead so much faster and more securely than earlier, less successful ones, that devoted immense intellectual skill to building on inadequately tested axioms and practical experience.

These processes cannot themselves establish strictly and positively any law at all, or distinguish between a chance pattern of events, a system bound by causative laws, or any of the many conceivable permutations and combinations of mixtures of regular and random happenings. They can only help to identify laws that are believed to exist.

Probability is no help whatsoever in this matter. Judgement must now be made after the events that determined whatever pattern exists. It depends on them for the presence of the observer, and his ability to wonder how and why they happened. So the question of how improbable they were before they occured is quite irrelevant, as well as unanswerable.

Scientists, like Christians, are denied any irrefutable mechanical sign, and forced to base their lives on personal experience and faith.

Not even in its analysis of the physical aspects of the world has science been able to create a self-sufficient system that can stand on observation and reason alone, without recourse to prior intuitive beliefs. Once these have been accepted, as a mere intellectual game much of the system can be operated without reference to them. Yet the feasibility of maintaining this pose for any length of time outide the most secluded backwaters is questionable; while the choice of assumptions and methods then becomes quite capricious. This may be magnificent; it is certainly not scientific.

The scientist who imagines that he has purified his work of all presuppositions by rejecting the original Christian basis for his customs has done nothing of the sort. He has simply endowed sub-sections of that code, the parts defining the unity and rationality of the material universe and the rigour and integrity of scientific analysis, with independent supernatural authority. Then, unless he abandons all interest in topics outside his own speciality, a stance of doubtful feasibility, he has to believe that these are capable of dealing with aesthetic and social problems as well. Yet the instruments used in natural science are manifestly incapable of even recording many of the elements and relationships experienced in these fields.

Logical analysis and physical measurement alone cannot conjure up ideas and existence out of nothing. Without an initial belief that there is something to be examined and some notions from which the means of doing so can be developed, the operations of science cannot start at all. Scientific methods are an extremely powerful way of gaining further knowledge, of refining ideas, and of testing how well any of the latter fit the former. Yet they themselves depend on the instinctive belief that the physical senses detect something real, and on the validity of human imagination as a source of concepts by which observations can be rationalised and understood.

The scientist's urge to try to attain absolute certainty by

rigorous deductions from completely controlled experiments is in many ways admirable, but in fact inherently impossible. It is the modern successor of other romantic quests, such as the Holy Grail or the Universal Solvent. The feat of lifting oneself by one's own bootstraps is very flattering, and deeply engrained in much folk-lore. Yet there is a simple mechanical explanation of why this operation is in reality impossible. So attempts to carry it out only lead to inelegant and uncomfortable conclusions.

Relationships

Modern science is not simply the sum of a vast number of individual attainments. It requires a massive network of co-operation, running vertically in that today's latest triumphs are based on the accumulated information, wisdom and skill passed on by previous generations, and horizontally in the need to draw on large and ever growing circles of contributors in many developments. The known universe is now too big and complex for any polymath to comprehend any significant portion of it. Only a well organised society with a large, efficient, honest memory can cope with the immense, diverse, but highly interdependent problems it poses.

Such a system is difficult and costly to run. So some polarising influence is needed for accepting the conditions and pattern of relationships that make it feasible. Some incentive is required for maintaining the high degree of personal commitment necessary to prevent it from breaking down.

When science moves out of the laboratory, and the very artificial conditions sometimes maintainable therein, the impossibility of the system functioning without some extra-scientific justification becomes even more apparent. Moreover, while a degree of detachment from other people and influences is often desirable in conducting particular investigations, sometimes a very high degree, complete isolation of scientific thought and activity from that of its neighbours is not now possible. Science has become so complex, and demands so much time and material for its pursuit, that its existence can only be sustained by highly organised communities. It is simply not viable in isolated 'Cartesian' enclosures.

Many whose support is essential would not grant to the acquisition of knowledge for its own sake and the strictly logical analysis of events the pre-eminent value accorded to them by scientists. Most would allow aesthetic or emotional, moral or religious factors greater weight. Yet all constituents of the community must at least tolerate one another, if they are not to destroy their life support system by tearing it apart.

Science, as a sub-group dependent on the main group for its own survival and freedom, cannot afford to contravene standards of conduct that are indispensable for the health of the whole community. It must be be an organ contributing to the well-being of the body politic, not a cancer. Equally science, being an essential, constituent part of society and not an intruder, is fully entitled to criticise and make constructive proposals for the development of satisfactory codes for governing the whole. Yet it must ensure that those more detailed ones it adopts in its own work are reconcilable with those held by other sectors, whether closely or remotely related.

The degree of specialisation now common in all walks of life makes it difficult to recognise this obligation to set one's own concerns in a wider, indeed comprehensive philosophy. It allows side interests to atrophy. Contact with ideas and experience in other fields of human activity then becomes so attenuated that they seem irrelevant. Yet it is misleading to confuse such concentration with scientific rigour. This epithet can only be legitimately applied to those who are always conscious of factors that have had to be ignored for a time, to enable some facets of the universe to be investigated as thoroughly as possible.

While some isolation is tolerable to facilitate the experiments and analysis feasible with the resources currently available, it is absurd to imagine that formulae which adequately rationalise observations so made can always be applied outside the aseptic environment within which they were obtained without modification. Careful consideration must be given to whether anything excluded may change the situation so completely that they are useless, without radical revision. Even if the expanding territory of any subject does not soon reintroduce some awkward factors that have been omitted, opening the door of the laboratory or office admits a host of them.

Crude, irrational methods of making choices are quite incapable of controlling the scientific complex. So if those who especially value reason, and wish to promote its use, behave in a patently irrational fashion, they must not be surprised if members of other groups almost revel in deciding matters of policy by emotion or caprice, even in technical affairs. Just as some common foundation of belief is essential for the development of physical science, if the scientist is to obtain the cooperation he needs from a lot of his neighbours, the creed he adopts must be one that is able to provide all with a modicum of common purpose.

There is no loss of any real measure of freedom in recognising the inherent limitations of one's own discipline, and its dependence on society for any existence at all. Any higher degree is self-destructive anarchy. The scientist trying to maintain absolute independence is behaving like a child sulking outside its own home, because other members of the family have tried to throw their weight about too boisterously.

Such self-centred self-pity and unreason is more dangerous to the scientific professions than to others that are not so deeply committed to enlightenment through reason. Accepting their social responsibilities would put scientists in a better position to influence those of their brethren addicted to less rigorous disciplines, or no discipline at all, and help to curb prejudices that threaten to divert the achievements of science to global slavery or destruction. Flexibility ceases to be a virtue if it is carried to the point where lack of sufficiently secure attachment between members allows the whole structure to collapse.

Chapter 6

Viability

The first two assumptions can be maintained, albeit somewhat untidily, without much concern for what is happening outside the laboratory. The third presents some awkward problems to the scientific isolationist.

For the pursuit of 'pure' science, of knowledge for its own sake, is no longer honestly practicable outside the narrow confines and meagre resources of a hermit's cell and cabbage patch.

It never was more than an unapproachable abstraction, like the Ideal Heat Engine. That it was attainable was an aberration fostered by the inordinate wealth of nineteenth century European aristocracy. If it is allowed to linger as a pointer to an unattainable perfection of pure learning, the slightest taint of 'regardless of the community' destroys its legitimacy. Criteria from some system that recognises and balances the interests of all involved is essential for resources to be directed intelligently, and the benefits distributed equitably and rationally enough to preserve science and the society that supports it alike.

In these circumstances, if the scientist is to maintain a truly scientific stance, he must ensure that even the social conventions he adopts do fit observed social facts.

Social stability

Records show that every community that has ever existed has found it necessary to impose some controls for securing a viable level of social cohesion. These have always been related to some notions of the rights of individual persons and the function of the group. Such concepts cannot be extracted from any regularities noticed among the interactions of swarms of insensate particles, regardless of whether these seem completely pre-determined or appear to have some indeterminacy. Only what is or is not can be detected by observation and analysis. Any judgement of what ought or ought not to be done requires some external criterion.

The simple fact that no individual or society exists without some beliefs about these matters, and the impossibility of importing any order into social relationships without reference to some such authority, has been too easily and generally disregarded. This may be due to its having been presented in terms that offend the strong strand of Stoicism that is deeply woven into the traditions that have played a major part in developing modern science.

Without a belief in personality transcending the mere physical reactions in and between masses of inanimate particles that happen to have assembled in configurations that are classed as 'people', cries of human ecstasy or pain are as meaningless as the squeal of brakes. This is true enough; but it appeals to emotions that are unfashionable in dominant intellectual circles. Yet without grounds for thinking that ideas are meaningful and communicable, and some metaphysical criterion for judging their significance and value, there is no reason for preferring the kindly and rational ones of an Albert Einstein to the bloody and dramatic ones of an Adolf Hitler. Both are no more than the peculiar properties of two similarly shaped physical objects, emanations from the physical reactions occurring in particular assemblies of atoms.

The transactions of the proudest academies would be simply events, of no more importance than the decay of sewage. Science is ultimately as dependent as religion on an 'inner light' for its system of values.

Further, any scientifically adequate creed must fit the observed behaviour of human beings. The history of the present century has reinforced the evidence from many previous ones. Humanism by itself is not enough to maintain humane standards of conduct. Without some stiffening and polarising influence, the inborn selfishness of man mars or destroys the blessings that his potential and virtues should command. Sheer bestiality to their fellow men has reached its grimmest peaks in some of the most civilised nations. This patent defect, and the hopelessness of life if there is no power available to overcome it, are features that no scientific account of the nature of man can ignore.

The scientist may consider some forms in which the Christian doctrines of sin and evil have been cast rather grotesque. His own theories are even more defective, if they fail to cover some of the most manifest facts of life at all.

Scientists have made the written exposition of information and ideas the basis of their own system of communication and assessment. So it is odd to find widespread insensitivity to the limitations under which the Scriptures had to be composed going hand in hand with enthusiasm for primitive styles in other arts among them.

Many of the scientist's own formulae and pictures are much less adequate for describing the essence of the phenomena and relationships observed than are even the earliest religious writings for describing the relationship between God and man, despite having to deal with much simpler situations. Acquaintance with how scientific views have changed during the last century should convince anyone that our present conceptions of the structure of matter will appear cruder approximations to the underlying reality than the stories of Genesis after another couple of millennia, if anyone remains and bothers to read them.

The scientist should recognise similarities of outlook hidden by differences of technical jargon, instead of concentrating attention of the absurd claims made by extreme sections of the religious fraternity. His own profession suffers from similar blots. Although annoying, these exhibitions are no excuse for allowing himself to be put off likely openings for co-operative endeavour and research, and for abandoning any attempt to base his operations on an integrated framework of ideas that fits the full range of human experience, in social as well as physical studies. He might then notice how close some anthropologists and psychologists are drawing to Adam in their beliefs about the state of our ancestors, and their influence on our personalities today.

The manifest failings of the individual human being have led some philosophers to try to found their humanism on a communal base (family, tribe, nation, race, church, state, party, class, etc., etc.). This can only be done if one ignores the fact that the most inhuman behaviour has been practised by the most highly socialised societies. The mob is notoriously even more irrational, fickle, and savage than the individual.

Social pressure can only be helpful in maintaining the integrity of personal conduct on which the stability of society and the progress of science alike depend in those golden periods, usually short and widely spaced, when the sacrifices of the forerunners and a fortunate conjunction of circumstances allow the highest ideals to dominate the aspirations of a community, rather than the lowest or some mean compromise. When the right course of action happens to fit the fashion of the moment, no problem arises. It is when they conflict that tensions occur.

When following the line of duty becomes unpopular as well as against the interests of the individual, no group pressure will support the maintenance of sound standards. Any communal influence acts in the opposite direction. Only complete personal conviction of the obligation as a matter of conscience, strengthened by the belief that one serves some better cause than mere self-satisfaction, can be effective in resisting the many internal and external pressures for the easy rather than the right solution.

So any honest, adequate scientific creed must include components capable of sustaining both the individual conscience and a tradition of integrity in any community. The apparent weakness of both should logically leave them impotent before the sheer weight of the forces opposed to them.

The proper functioning of the scientific system demands the prior acceptance by those involved in it of certain attitudes of mind and a pattern of behaviour that cannot be justified by the same methods as propositions made within its own territory. So there is an obvious need to set out these ground rules clearly, so that all can recognise them, assess them critically, and assent to them if they want to harvest their fruit.

Initially these were provided by the common Christian heritage of Europe at the Renaissance. The failure of later generations to notice the connection is a classic example of the dog's existence and evidence being ignored, because the dog did nothing but be there and behave in its normal, natural way.

'Western' science misleads itself and others by its pretence that it works without any metascientific dogma to support it, unlike others such as Marxism, which make no bones about the importance of their creed. The real difference is that it does not acknowledge the source and implications of its beliefs.

This pose makes the testing, let alone any hope of reconciling, competing systems impossible. Yet unless this is done, friction between them may well precipitate conflict of an intensity capable of exterminating human existence. If a responsible dialogue is to take place, out of which mutual respect and understanding might grow, both parties must agree that there is something about which to talk.

Isolationism

Some scientists are today in as great a danger of isolating themselves from reality in their laboratories as their priestly counterparts are in their sanctuaries. Members of each sect tend to place a halo of infallibility around the particular system of codifying and rationalising experience to which their temperaments and early training have committed them. Yet one of the features that has contributed most to the success of modern science is that it recognises and accepts the defectiveness of human nature as it is. So it adopts techniques that can in time correct the inevitable mistakes, if they are applied honestly and consistently.

Each group needs both freedom of inquiry, and a discipline to prevent its intellectual efforts from degenerating into the fruitless multiplication of irrational taboos. Each must establish a *modus vivendi* with the community it serves, and on which it depends for services and survival.

Each must, therefore, ensure that the beliefs and concomitant codes of conduct by which it regulates both its internal operations and its external relationships are reconcilable with the others, and with those of many co-operating or competing groups. Neither can adopt different standards according to which way it faces, or it will rightly be dismissed as an intolerable bunch of hypocrites.

The barriers to re-integrating our culture are neither reasonable, aesthetically attractive, nor instinctively sound. They are indolence, leading to indifference to the dangerous social tensions that failure to do this entails, and pride, resisting any infringement of a completely mythical self-sufficiency.

There will always be individuals or sections of the community that reject the need to relate different spheres of experience to one another in a coherent fashion. Limiting their fields of interest, and the scope of their methods of inquiry is much easier. It also avoids the danger of exposing vulnerable aspects of their behaviour to examination that might prove embarrassing.

Yet that decision is itself an act of faith. The set of norms, however oddly chosen, by which they guide their reactions to any situation is a creed. Those who prefer to seek a more unified pattern of responses should not allow the sometimes conflicting jargons current in different specialities to obscure the meaning or function behind them. It is the substance within, not the label on the bottle that matters.

Reconciliation

There is no real difference between religious myths and scientific theories. Both are generalisations designed to clarify observed facts and relationships. The elevation of either to the status of laws depends on the amount of information available for testing them, and the closeness of fit between theory and practical experience.

There is no more reason for rejecting the contribution of religious prophets to the solution of moral problems than there is for refusing to use complex numbers in mechanical ones. Both are insights to which the contemplation of different aspects of human experience have led outstanding personalities. As the introduction of these concepts leads to the integration of wider ranges of phenomena, and provides the means of solving problems that simpler techniques cannot handle, the honest scientist must either accept or reject the use of such devices in all types of investigation.

If he admits them, the sole criterion of value must be how effectively that help to elucidate the issues to which they are applied. To attempt to segregate the over-lapping and interrelated forms of human experience by uncrossable barriers, and so accept methods as legitimate in one subject but disallow them in the next, is quite arbitrary and irrational. St Patrick's shamrock analogue of the Trinity is as accurate or inadequate, true or false, helpful or misleading, as an Argand diagram representation of complex numbers. The Russian cosmonauts who did not see God did not see the force of gravity either. Fortunately for them, both were there all the time.

Ideas about many things and relationships are continuously being modified and refined. Meanwhile they may be difficult to record or visualise by means of established symbolism. This does not invalidate the belief that something exists that is worthy of attention, the attributes of which are slowly being revealed.

Every individual holds some moral standards. There is even something that they dare to call 'honour' among thieves. All groups find that some common moral code is essential for their health and existence. So there is no reason to doubt that there is just as much reality behind this form of experience as there is behind physical experiments. It too will repay deeper investigation by yielding a clearer understanding of the laws that bind the entire universe.

Science must be able to defend its right to conduct strict but unfettered inquiries (NOT experiments, if the evidence suggests that these may be unduly destructive, that the likelihood of loss or injury greatly outweighs that of gain) in every field of knowledge. So it must play its full part in analysing social experience, in the hope of discovering a set of behavioural standards that includes such freedom as one of its components. Only if such a system exists is there any rational ground for hoping to retain personal and professional liberty in the highly organised structures that technological progress demands, and in the intensely concentrated communities that it tends to create.

Of engineering

If even natural science needs some coherent groundwork of belief on which to base its methods of operation, and the codes of conduct that enable it to continue to work effectively, such a requirement is manifestly even more essential for engineering. For the art is committed not only to applying science, but to applying it for the welfare of mankind. This operation clearly requires some set of values by which to measure whether any project seems likely to yield positive or negative returns to its beneficiaries at the design stage, and to confirm whether it has done so when it is completed. Unless some means exists by which these tests can be carried out, the performance promised in the prospectus cannot take place.

So a profession dedicated by its charters and traditions to directing natural science for the use and convenience of man has a duty to take a lead in maintaining a radical ethos in the community it serves. Today society is losing contact with reality behind an almost impenetrable array of gadgets and preformed attitudes. These distract attention from fundamental issues, and obscure them.

If engineers are to follow the example of their accredited heroes in the Renaissance and the Industrial Revolution faithfully, they must make their philosophy conform to the evidence, instead of following the Schoolmen in subjecting facts to dogma. Indications that reason by itself is not enough must be accepted. The popular search for 'fool-proof' mechanical systems is the most blatant folly. The vital dangers lie not in the technology, but in the way man uses it.

The layman must often find it difficult to appreciate the empirical way in which any engineer or applied scientist decides which of the many instruments available to him he will use in trying to find the best solution to any problem. So the former is reluctant to grant the latter the degree of confidence in his judgement that he considers the due reward for his qualifications and experience. Yet trust is essential for adequate co-operation between patron and adviser.

At one moment the engineer is recommending caution, and further checks to prevent current techniques being pushed beyond their legitimate spheres of application into regions where errors, perhaps inevitably undetected in previous exercises, or factors omitted, rightly so in other cases, may become significant, or even dominant. The next he is advocating taking risks by pushing theory and technique to the very limits.

His choice should never be arbitrary, as it should always be defendable in rational terms; but it must often be intuitive, as in most creative work. Existing knowledge and skill alone are seldom enough. Precedents are frequently too remote or thin on the ground to offer clear guidance. The available analytical machinery may be too cumbersome to yield results in time, or a unique solution.

These difficulties are not peculiar to engineering. They are common in every field where rapid advances are being made. Yet they are particularly serious and trying in situations where a wide range of physical influences and a large number of groups of people are involved. In these, professional judgement is often the only means of making decisions while they still have any utility. For time is usually of the essence of any project.

The lack of sufficient time and resources for demonstrating with absolute, scientific rigour the validity of every stage of the reasoning does not by itself invalidate the process. The basic criterion applied, that there is adequate evidence in closely enough related cases to warrant the value of the method used in that case, is as sound an approach as possible, even if it is a very difficult one to check. Projecting ideas into ill-charted regions or the future requires imagination and taking risks, however carefully these are controlled by the scientific techniques and practical experience currently available. The infallible system is a gambler's daydream, not an attainable reality.

The task of integrating these onerous, sometimes conflicting imperatives often baffles members of the profession, patrons and outside observers. Doubts about the feasibility of the exercise can lead to a serious loss of credibility, unless the confusion can be resolved within the framework of some higher discipline. Grounds for assurance that the apparent contradictions can be reconciled are essential if society and the technology on which it depends are to retain sufficient coherence to survive. The conditions entailed by their adoption must not only be recognised, they must be satisfied, and seen to be respected.

Ideas must be capable of providing the axioms for a tolerably agreed vision of the universe. Ideals must promote a set of values and pattern of relationships through which service to the community can be offered with full consideration for the freedom and integrity of all. These are essential if engineering is to be able to discharge its specified function. And if science in general is not to be forced into an isolation that will in time be fatal to its own survival, even if its immediate form is being placed on a slightly elevated pedestal where the sacred cow can be milked to death.

The answer to the question posed by the title is that it is NOT an alternative. The distinction only exists in an artificial world. However handy it may be for classification purposes, it only records a single facet of the subject in each epithet. In the real world, all facets must work together if they are to be effective. Engineering must be BOTH a mechanical and a moral science if it is to direct the great power of natural science successfully for the use and convenience of all mankind.

Chapter 7

Necessary conditions

Mathematics is the basic language of science. Its processes and symbolism permeate every scientific discipline. Many years ago Gödel demonstrated that it could not be rendered self-sufficient. It requires belief in a higher order meta-mathematics to justify its axioms and methods.

So it is odd that there is still widespread reluctance to recognize that the problem runs under every discipline in the scientific hierarchy. If it is ignored, it is not surprising that the ground rumbles under one's feet. It becomes progressively more serious as the complexity of any subject grows. So its resolution is quite vital when attempting to apply natural science for the benefit of society, or conduct social studies scientifically.

The real question facing those committed to the development or application of scientific methods in relation to systems of belief is not whether the latter are necessary. It is which to adopt.

In searching for the answer, an answer if more than one is adequate, if so the best, both the mechanic and the moralist are entitled to the same privileges, and bound by the same rules. Each uses imagination for creating and projecting ideas. Each relies on insight and instinctive sympathy for assessing methods and standards. Each must submit to the discipline of observation and reason. Each needs to utilise robust, empirical ways of testing theories if anything worthwhile is to emerge. Neither can expect a higher order of correlation between theory and practice in the other's sphere than his own.

If a coin came down heads 1,000 times in succession, not even the strictest scientist would write to *Nature* to record his singular experience. He would have decided much earlier that the coin was loaded; and either thrown it away, or set out to discover how, and why. There is no valid reason for him to adopt any less realistic an approach to whatever moral beliefs he chooses to follow. Any attempt to apply double standards would forfeit respect for his own professional integrity, and undermine the security of his own discipline.

Modern science was based on Christian beliefs, whether openly or implicitly, even surreptitiously admitted. Its relevance, or that of any alternative propounded, to the continuing success of its operations and the way it is applied, must be evaluated in this manner.

One can in theory concoct a science to match any religion, or religion to match any science. Yet if any creed is to provide a firm foundation for modern science in practice, it must give authority to those articles of belief that the system cannot create or verify by its own techniques. These are vital both to its own internal working, and to keeping its relationships with other sectors of society in working order. Six articles are essential:

The *first* is that the natural world and all that is therein forms a coherent, rational whole, a universe in the scientific meaning of the term. If it is not, there is no reason for expecting to be able to find any laws valid throughout space, time, and every constituent discipline.

The *second* is that this entity is open to mankind's understanding, provided it is approached with integrity, humility and patience, that the external world is not a figment of human imagination. Unless it is, the whole structure of science is but a vain, childish fantasy.

The *third* is that the methods used instinctively for gaining some insight into the laws of nature are reliable. The techniques, physical and intellectual, used to collect, collate and test data and draw conclusions therefrom must neither be being queered by some external agency playing tricks, nor suffer some random or systematic internal defects. If they are not so trustworthy, no system of making, classifying and analysing human observations can lead to concepts legitimately describable as laws.

The *fourth* is some idea of the nature of mankind that will provide a basis for the high level of personal responsibility and of co-operative relationships that the system demands to work properly. These are costly and difficult to maintain over long periods, so some strong incentive for doing so is necessary.

The *fifth* is some standard by which to direct and evaluate actions. Without reliable criteria for measurement and orientation, concepts such as 'progress, benefit, utility, convenience, welfare' are meaningless; and anyone starting with what he thinks is a clear idea of them can find his vision becoming dangerously confused.

The *sixth* is some rational ground for hoping that mankind can maintain and measure up to these reasonably well. The long march to the New Jerusalem is a hard, tiring slog over most of the route, which can tax even the best disciplined and equipped formations.

All these are essential if the system is to work. Yet some are very vulnerable to being taken for granted in highly developed communities. Then they are in time overlooked or ignored. Finally they are abandoned or rejected, with calamitous consequences.

Many problems can now be solved by well established precedents and widely distributed secondary standards. So the need to relate them all coherently through some fundamental set of principles is often forgotten.

The meta-stability that good habits can confer if they are correctly set in the first instance obscures the narrowness of the safe path, and hence the difficulty of neither despising the natural world nor allowing material considerations to dominate society. Worshipping an unknown god is still a path that leads to superstition and despair. Deifying man is patently absurd.

Subtle new variations of selfishness and barbarity are constantly being devised. False trails must be identified, the bench marks needed for this purpose must be refurbished and pushed forward just as regularly. Any generation that fails to discharge these duties is living on and dissipating the capital built up and bequeathed to it by its predecessors, regardless of its own steady impoverishment, and the blighted heritage it will pass on to its children, if any.

The vital question is not the nature of the art of engineering. It is under what authority it can best hope to satisfy the professional aspirations of those who practise it and the social expectations of those who patronise it.

UNDER WHICH KING, BEZONIANS?

Chapter 8

Jesus of Nazareth?

Some infrastructure of belief is necessary for coherence of personal and social behaviour, if a technologically sustained society is to function efficiently enough to survive. The present system evolved in a society whose processes of thought and codes of conduct had been moulded by the Christian religion. This does not necessarily entail that it is still essential. Nor even that it is the most convenient basis on which to operate now.

Disposable creeds

It is arguable that the beliefs from which the propositions flowed which inspired those who pioneered the system can be abandoned as irrelevant. That the secondary standards derived from them can be maintained by their utility, and force of habit. That the unity and rationality of the universe, and the need for high standards of personal integrity and co-operation amongst those who would understand and profit from knowledge of its laws are manifest. That criteria for judging what is good for mankind to guide applications in practice are self-evident.

Alternatively, it may be suggested that the original traditions are only of value in childhood, as a means of evoking a number of ideas and responses useful in later life. That once acquired these can be retained independently. So the obligations of the creed from which they sprang can be discarded on reaching adult status. This attitude seems common today. Schools are expected to instil decent standards of conduct into their pupils despite the indifference or conflicting example of parents.

The persistence of such attemps to have and to hold the advantages of a particular system of belief without accepting the responsibilities that are the other side of the coin is odd. In theory it is neither possible nor honest. In practice there is ample proof, in the retreat from civilised behaviour, of the futility and dangers of such hypocrisy.

Any creed will do

Even when the need for some continuing metaphysical support is admitted, it is often suggested that a better set of axioms is available. Or that any of a whole series of sets is adequate, as long as it is self-consistent and includes certain elements. Such doubts are naturally raised by the great contributions made to science by men of other faiths. Many Jewish, Islamic and Marxist examples spring readily to mind. These have at times in some sectors excelled the achievements of Christianised societies contemporary with them.

This is as misguided as suggesting that the calorific theory of heat is as good as the kinetic one, because some thermodynamic problems can be solved by treating heat like a fluid. Tolerably accurate fitting for some purposes can be achieved when only partial congruence exists between the actual situation and a model used to analyse it. No progress would ever have been made if some problems had not been solvable despite erroneous ideas about many facets of them. Indeed, initially about most. Without this degree of tolerance in the way the universe functions, the gradual processes of evolution could not work. That they do is even more clearly demonstrated by the history of human understanding than by the record of physical development.

There are elements common to Christianity and some other faiths. Often the failure of Christians to honour the full implications of their creed has in fact denied the sincerity of their commitment to it. It is not surprising if others applying the common parts more wholeheartedly have then achieved more. Particularly in those spheres where those elements disparaged by Christians are the most important factors.

Back to basics

But the close connection between Christianity and the most powerful system for acquiring knowledge yet seen on Earth makes it the logical starting point for seeking a firm foundation for a technologically armed society. Its ability to provide the axioms that science needs but cannot discover or justify by its own methods must be tested. Comparison of its effectiveness with alternatives on offer will give an indication of whether a continuing connection is necessary for the maintenance of the system in healthy condition.

The exercise may also throw some light on whether there is any intelligent ground for hoping to retain ideals such as personal freedom and human welfare. For the highest development of these has been nurtured by the same source.

Before doing so, however, one must emphasise that the

demonstration of close correlation between the scientific system and Christian doctrines does not and cannot provide irrefutable proof of the truth of the Christian revelation. Untold harm has been done by Christians ignoring their Lord's clear warning that no such mechanical sign would be provided. Too many try to reduce a personal relationship dependent on practical experience and faith to a set formulae and rules.

What it can do is help to clear some reputedly scientific obstacles out of the way of people who realise the inadequacy of other ways of getting to grips with some basic, unsolved, otherwise insoluble problems. They can then test the efficacy of Christian discipleship for themselves, free of irrational inhibitions.

Shared features

Most monotheistic religions offer justification for the first article. If there is one God who made and governs the universe, and who is himself rational and consistent, the belief that the natural world must have the same characteristics is but the recognition of the imprint of his nature on his creation. Nor is this belief incompatible with the belief that he has given to one section of animate creation the freedom to follow the light that he has implanted in them, or ignore it and try to snuff it out. Christians must insist on the unique significance of the revelation of the loving action of God to which they witness. Yet to deny that all religions which believe that one and only one God exists must worship the same God, however different their vision of his nature may be, reduces their own conception from a truly comprehensive one to a tribal deity. Indolence or the influence of pagan relics have often reduced Christian practice to decadent, polytheistic forms. It is not then surprising that others who have kept the absolute unity of God more firmly in the centre of their vision and lives have contributed more to the growth of true knowledge.

Moreover the acknowledgement of one sovereign creator imposes the humility in conducting inquiries and affairs that is essential if they are to be kept on the right lines. For this requires the recognition of the limits as well as the potentialities of investigators and instruments. Although the mirage of the definitive, final break-through still lingers in the sub-conscious of some romantics, every genuine advance in fact opens the way to asking more questions than it answers. Whatever the state of the physical universe, the realm of the human mind is continually expanding. Safe progress can only be made in such circumstances if ignorance is as frankly admitted as knowledge.

However unfashionable it may be in some quarters, caution in the use of power, social as well as mechanical, offers greater rewards if all forms of human satisfaction are included. Understanding oneself and respecting one's environment leaves fewer memorials of their forebears' intemperance to any posterity.

Such a belief also provides a motive for adopting the pattern of social behaviour that is necessary if mechanical science is to contribute to the use and convenience of mankind. Responsible stewardship of the talents and resources granted by the Lord of the universe to his creatures is likely to be demanded. Co-operation among his servants in pursuit of his purposes is a similar obligation.

Yet the assumptions that give validity to the methods and direction of scientific inquiry require more than just singlemindedness in creation. They depend on the nature of the creator, and his attitude to his creation. He must not be indifferent or hostile to his creatures, or a mere prankster, if the assumption that the insight they think they are obtaining into the workings of universe is to be true. So deeper knowledge of his character, and reliable information about his purposes are still needed. These are often provided by a combination of sacred writings and the edicts of a priestly caste. Incentives are frequently postponed until the hereafter.

Here again there are some features that are common to Christianity and other monotheistic religions. But it is in these areas that Christianity injects the factors that have made Christian science so successful. For it offers a unique view of the essential nature of God, and of his relationship with mankind. It is these that stimulate the responses that make the system work.

Unique features

Any transcendent God who wants his creatures to develop in his preferred way can bind them tightly with constraints that prevent them erring. Alternatively, if he wants them to follow it freely, he must provide some revelation of his purpose that enables them to understand and respect it. This must be cast in a comprehensible form to attract a response from the natures that he has created.

In Christianity both the authority demanding and the incentive encourging this response stem from the belief that man's relationship to God is most accurately described as that of a son to a loving father, if man will accept the offer. By basing God's entitlement to obedience on reaction to his love, rather than on fear of his purity and power, or wonder at the immensity and complexity of his works, it presents the relationship in a well known, universally respected form. Moreover it is one that can generate the depth of personal devotion that is needed to make science self-sustaining and self-correcting. It also offers an antidote to the despair that can easily result from contemplation of the other aspects of his nature alone.

If this belief is valid, instinct and experience are reliable methods of trying to understand God's creation. They carry his Son's assurance that one can indeed judge events and relationships by the innate, God-given sense of the value of their effects. Freedom of inquiry is the natural privilege of heirs. The obligation to love and serve one's fellow human beings is manifestly included; since all men are brothers. The whole pattern of integrity and respect on which the effectiveness of modern science depends evolved quite naturally and directly from this simple, familiar and highly regarded relationship. So the success of this creed in creating the conditions necessary for the improvement in man's understanding of the natural world and the marked deepening of his social conscience was not surprising.

Moreover, the standard by which we are bidden to guide and measure our lives is a real, live person. This offers a much more attractive and effective norm than a set of rules and regulations, and a more comprehensible one that any academic or mystical abstraction. These tend to be too remote from daily experience for the vast majority to see their relevance to their ordinary working lives. Their claims seem too arbitrary to command wholehearted loyalty, instead of a minimal, grudging compliance under duress that can readily be withdrawn. Yet a general diffusion of an adequate level of understanding of its basic creed is essential if any society is to function properly.

Attachment to one who has mastered an art is for most a surer way of becoming imbued with the spirit in which its problems should be approached than studying a textbook. Shared experience enables the novice to assimilate his tutor's skill in selecting, testing, and using tools and formulae easily, and acquire the judgment needed to discount many personal and transient factors that tend to distort his judgment.

That the creation of such a direct, personal relationship between God and every individual human being was the primary object of the consummation on the Cross of the whole life and teaching of Jesus is often obscured by the properly central role of the Bible in all Christian activity. This book is too often treated as a legal manual. Yet its claim to be the 'Word of God' stands simply and squarely on its unchallengeable position as the earliest and most reliable evidence — the best in the scientific sense, despite the failings of its human authors, editors and translators — for coming to know the 'Word made flesh'.

Exposition of the facts of life in this form was not only most expedient, it was probably essential. Nothing short of the compelling, concrete manifestation of the self-sacrificing love of God shown in the life, death and resurrection of Jesus was capable of making it a reality to most human beings, and freeing the concept from every limitation imposed by the self-centred nature of man. It was the critical experiment needed to demonstrate a love otherwise beyond human comprehension. It established its feasibility in the face of deep, sincere and reasonable human doubts.

It was also convincing proof of the Father's determination to provide a means of reconciliation with his erring creation whatever the cost, to reintegrate a relationship broken by the wilfulness of his children.

The tangible expression had to be set in a real, historial context. Yet the boundless power of the Spirit of such a loving God could not but be free to speak directly to men and women of all times and in all places. Events of this magnitude and significance cannot be contained within the physical limits of space-time.

Other creeds

Some of these try to obtain similar results by talking about the brotherhood of man, while rejecting the fatherhood of God as blasphemous or ridiculous. Yet brotherhood means nothing but that two persons are related through a common parent. Trying to separate two facets of a single concept is as futile as imagining that the clever technical feat of slicing the old pound note into two yielded two ten shilling ones. The result was waste paper of no value whatsoever.

True brotherly love does not depend on similarity, whether of inherited or induced characteristics. Indeed the great virtue of the relationship is that it creates bonds of affection that transcend differences of character, interest and background. The fact that all human beings have the same number of chromosomes is no more relevant to the existence of mutual liabilities based on affinity than the colour of their hair, eyes or skin.

The only rational way of justifying the doctrine that all men are brothers other than the spiritual one expounded by Jesus is in following the author of Genesis, in postulating a single common ancestor, an Adam however protoplasmal and demythologised. Such a belief set in now known time-scales is more tenable than many critics of the Bible admit. So is the corollary that at some unknown point in its evolution the species passed a critical stage that entitled it to assume the description 'human', regardless of whether that step required further intervention of the Creator or not. Yet such a belief is too dilute, mechanical and remote to command any strong allegiance, even if it were universally adopted. Moreover, these two explanations are not incompatible.

Without some vision of human relationships that might hope to evoke the very deepest, universal reaction, the only other method of getting sufficient co-operation to retain control of the enormous powers that can now be released by applied science is for one arrogant, irrational dogma to impose its authority on or to eliminate its opponents. The frequency with which this solution has been tried and failed demonstrates its inability to deliver the goods. Now that so much destructive power is so widely distributed that any further exercise of this kind might lead to the extermination of the human race, it would seem sensible to abandon that line of experiments.

If, however, Christians allow selfish considerations to override the obligations of their faith, while other religions observe the practical implications of it more consistently, they should not be surprised if the latter achieve more than the former. Marx's religious background rightly revolted against the personal degradation tolerated by nominally Christian communities, so

Marxism not unnaturally took a materialistic form. Even then it may well inspire a more real sense of brotherhood than the Christianity from which it purloined all its good ideas and ideals, if the latter remains indifferent to the claims of those outside the very closest biological or geographical, political or intellectual connections.

Justification

Christianity offers the simplest comprehensive set of ideas about the nature of man and the universe that forms a coherent, rational unity capable of providing an environment in which science and society can flourish together. The idea of a vocation to follow a particular discipline only makes sense if there is someone to call. An authority is provided for an instinctive trust in empirical methods, and a freedom of inquiry only limited by regard for the interests and personalilites of others. It presents an unrivalled example of unlimited personal commitment as a standard to follow. These features are set in a matrix that offers hope for mankind, while recognizing its patent stupidity and inhumanity. Man is so patently flawed that he may use the power released by science to destroy himself. So if men are to continue to penetrate the mysteries of creation, there must be some component in their beliefs that suggests that whatever spirit animates the universe is ready and willing to suffer with and redeem its creation.

All these elements are essential if science and society are to make real progress, and avoid the danger of either destroying or enslaving themselves, by ignorance or folly, by despair or lack of mutual respect. No other creed provides all the necessary ingredients as concisely and conveniently. The unity of nature cannot stem from a polytheistic faith. Other monotheistic creeds cannot tolerate the Christian claim to sonship of God, nor therefore justify the relationships it implies. Non-theistic formulations either become pure abstractions of no practical utility, or else elevate some human being, usually the proponent, to an eminence he can neither rightly nor securely maintain.

If all competing creeds are treated equally rigorously, Christianity offers similar advantages to those for which the Copernican view of the solar system was preferred to the Ptolemaic, relativistic mechanics to classical. It provides a better frame of reference for fitting all observed events into a consistent, rational pattern.

Some people may hanker after change and excitement, some for a less demanding way of coping with the problems that life entails. There is no evidence that an alternative code exists that can provide a comprehensive and reliable foundation for all necessary beliefs. Nor can civilised man peel off the advantages that have sprung from adherence to a system of values based on Christianity and retain them, while rejecting the authority on which they depend. They flow from a source that links responsibilities to the blessings that it confers. Mankind cannot hope to enjoy a Christian environment without Christ.

No demonstration of congruence between a pattern of events and a set of principles by which they can be rationalised can establish the truth of the latter beyond all possible doubt. Only faith has the right dimensions and so scope for such an achievement.

Yet the evolution of modern science through societies moulded by Christian modes of thought and behaviour suggests that the principle of least action may be as relevant in determining the paths of social phenomena as those of physical ones. Such a thesis would also explain the failure of other highly endowed civilisations to create as effective a system for exploring and understanding the natural universe. The evidence is in favour, though not conclusive proof, of the truth of the Gospel rather than its falsity. Thus it provides as rational grounds — if this necessary condition for being a scientist were sufficient for being a Christian, which again it is NOT — for accepting the Christian faith as any scientific theory.

Chapter 9

In Trinity?

The importance of a monotheistic creed on which to base modern science's belief that the universe is a coherent unity seems to demand some comment on the doctrine of the Trinity. For this can appear to many, in all fairness is sometimes made to sound by some who subscribe to it, a tritheistic creed.

Closer examination of the prime evidence shows that this is a false impression. Reflection indicates that it stems from the persistence of absolute, classical modes of thought. The retention of such encumbrances among the scientific professions and their avowed allies is strange. For the developments in physics that earlier in the century resolved the anomalies that demonstrated the limitations of classical mechanics, and provided the impetus for the rapid advances that have since been made, turned on the recognition that the precise definitions that previous schools had assumed were eventually attainable were simply not feasible, even of the most elementary phenomena.

Whether light was corpuscular or wave motion was the arena for the scientific counterpart of the theological battles waged over '*homoousion*' and '*filioque*'. Fortunately scientists could not then attract political patronage for quite such vicious persecution of opponents. The outcome was acceptance by all that it enjoyed both natures. Of a wave, despite the absence of any identifiable medium that could transmit it. Of a particle, despite a tendency to materialise and disappear like the Cheshire Cat. The two aspects are reconciled in theory by a wave-like mathematical abstraction that describes the probability that a material-like particle will be found at any point in empty space. Which facet is more significant in practice depends on the experiment being conducted, and the instrument being used. Yet no one denies that light is a single, unitary phenomenon, just because the human mind has difficulty in visualising this relationship between two radically different concepts, and has to record it in very abstruse, complex language rather than simple, everyday terms.

At the other extreme from elementary particles, and particularly relevant as one of the key creedal words is involved, a similar indefiniteness of language is found to be irreducible.

For individuals behave as quite different persons in different situations. They instinctively assume different personalities for dealing with different relationships. Yet the fact that people exhibit different characteristics when reacting to different roles or circumstances does not invalidate their description as unit persons. Nor the belief that all facets of their personalities are integrated in some mysterious way, even if how is incomprehensible to everyone else.

Similar irreconcilabilities emerge whenever one probes deeply enough in trying to define as precisely as possible the nature of any natural phenomenon. Adequate descriptions have been shown to require language that allows some degree of indeterminacy. Only by such means can irrefutable evidence of apparently incompatible characteristics be reconciled. As with the human eye, there are limits to the resolving power of the human mind. The problem must be particularly serious when human beings try to express their insight into the nature of transcendental entities.

This predicament inevitably causes difficulties when people exchange experience and ideas, as is both their duty and their right, in trying to understand themselves and their environment better. Only particular facets of anything can be illustrated by methods that are immediately assimilable by the senses. The moving pictures of electrons whirling around multi-coloured packs of billiard balls that are regularly used to represent atoms and nuclear reactions, and the three dimensional lattices with multi-coloured nodal connectors that are used to illustrate the structure of complex molecules like DNA, are simplified presentations of but a single factor, the spatial relations between the units involved. Such scientific demonstrations are even more limited in their scope for conveying the full, true nature of the objects concerned than are many artistic impressions of God. They are nevertheless valuable teaching aids if they are used with discretion.

One way of dealing with difficult problems is to use more highly developed languages capable of encoding many facets of a situation simultaneously, such as mathematics. Having done so, it is nonsense to pretend that wave functions or space tensors are any more readily comprehensible by the man in the street than the Athanasian creed. The advantage that higher mathematics retains is that it has to be specifically taught. So if it is well taught, the significance and limitations of its symbols are expressly and continuously expounded. Hence the desirability of exposing a much higher proportion of the population to disciplines such as engineering. There both the strength and weakness of this tool, the need to recognise realistic limits and record them in any communication, and knowledge of how to use but not abuse it in practical affairs are essential parts of the curriculum.

Less can be done to discipline the kaleidoscopic variety of literary forms, even when used in scientific exposition, except to warn users constantly of the pitfalls. One of the minor, amusing disappointments when climbing on the nuclear band-wagon from older established branches of engineering was having to accept the impossibility of persuading the high priests of that recently separated sect that it was inconvenient to use the word 'flux' to describe a scalar function. In every other science it denotes a vector quantity. Inconvenient, awkward for outsiders, but not insuperable as long as one remembered which jargon one was using. What is important is that users of any variety should remember that words, like numbers in statistics, which may have quite clear cut meanings in some contexts, may suffer from a degree of fuzziness in others.

It is vital that whatever method of communication is being used in any situation, whether verbal or numerical, visual or oral, transient or permanent, its limitations should be recognised, admitted and recorded.

This was not done when the turnover of generations and the growth in numbers made it expedient for the Christian Church to try to reduce the direct confession of its original creed — Thou (Jesus of Nazareth) art the Christ, My Lord and My God — and immediate recognition of His Spirit at work to some formal, authoritative statement of the essence of the faith. The limitations of even the best language available at that time were not recognised. This basic defect was aggravated by the well-known tendency of institutional hierarchies to pile Ossa on Pelion, the length of creeds grew in almost direct proportion to time and numbers. The sophistry of the contemporary intellectual climate powered the next stage. The arrogance of the political order by which Christianity had been adopted as the state religion finally boosted some strange artifacts into orbit.

These conspired to obscure the naivety of human experience

and vocabulary. They encouraged theologians to try to impose a definitive structure on the Godhead. Such intemperance exceeded that of the engineers who had tried to storm Heaven with bricks and mortar at Babel. As with that earliest mythical project, they confused the issue and split the community. Yet repetitions continue throughout history, of man's arrogating to himself power that God laid aside when incarnating himself in order to restore broken relationships with his creation.

What was surprising in such circumstances was how good the result turned out to be. Except to believers who hold that the long tale of accidents that have preserved the faith through many translations between times and places, cultures and languages, were inspired. For what matters in practice is not their inevitable failure to define conclusively how the different aspects of God's being are related, but the ability to recognise them and their significance. No epithets, however well refined by human thought and experience, can hope to describe the full, true nature of God. Yet the important features were set out accurately. There is only one God. His different manifestations are best described as personalities. These are the only known entities between which love can properly be expressed as existing.

Where the Fathers and their many successors went wrong was in trying to freeze the Spirit's continuous deepening of mankind's understanding of Jesus' revelation of the nature and purpose of God within their own vision. This was aggravated by their intolerance in trying — in his of all names — to compel others to conform to their preferred formulation on pain on exile or death. Christians should have heeded St. Paul's advice to avoid 'vain babblings, and oppositions of science falsely so called', and followed the example of Job. He achieved reconciliation when he confessed that 'I uttered that I understood not; things too wonderful for me, which I knew not'. It is comforting for the scientist to recall that Job was not condemned for maintaining his integrity and asking some very awkward questions. Though he had to come to terms with the fact that he could not expect to understand God's ways fully, being only, truly human.

Like all manifestoes, creeds must be used with the utmost caution. Knowing the taste of good or evil, honey or butter is a better way of learning to choose the good and refuse the bad than relying on the labels. How God's different manifestations are internally related must always remain a mystery. But that the loving action of God in creation, in the atoning work of Jesus, and in the inspiration of the Holy Spirit all flow from the one and the same fatherly God is the testimony for which St. John had to endure his long martyrdom, and to which the life of the Church has witnessed ever since.

One can hope that, as in many other sectors of our now paper bound bureaucracies, some pruning of vain repetitions and party slogans might be achieved. Yet the use of time honoured formulae need not sit too heavily on scientific consciences. Reference to 'the music of the spheres' commits no one to a Ptolemaic cosmology. The vast majority of practical problems are conveniently and satisfactorily solved by Newtonian mechanics and Euclidean geometry. Indeed, regularly facing and reflecting on the value as well as the limitations of primitive expressions in any discipline can be a useful reminder to those tempted to offer novel universal panaceas, a timely warning of the damage that can be done by uncharitable dogmatism.

Theologians are not the only group that are sometimes reluctant to admit the truth, particularly when it is simply that they do not know.

Chapter 10

Not that man!

Yet if Christianity was the source of the essential beliefs and the codes of behaviour on which the modern scientific system was founded, and still provides the most straightforward justification for retaining them, it should be as readily and almost universally accepted as gravitation, electricity or the roundness of the Earth. The absence of general recognition is contrary evidence that cannot be ignored. Some explanation is necessary of why so many no longer see the relevance of Christian beliefs to their personal, technical and social problems, and the relationship between them. Or ignore it if they do.

Throughout any such investigation, it is important to remember that Christians are nevertheless human. So phenomena that occur in human societies generally are bound to surface in their organisations. Indeed, as they are promised fuller life, these effects may be intensified. Yet they are called to transform society, not to conform to its defects. So the warning of the need for constant vigilance was given to them, and lies even more heavily on their Church than on secular organisations.

Penalties of success

Paradoxical as it may sound, one of the main causes of its present decline in advanced cultures is the very success that this creed achieved in dominating the group of nations that begot modern science. Any movement suffers from similar troubles if it attains a similar measure of acceptance. Science itself is showing signs of failing to hold the interest of the young and the loyalty of the community in general, despite its vital importance to the maintenance of the economic structure that supports the standard of living and the degree of freedom that they now demand and enjoy. Three consequences of success contribute to this regression.

Affluence

Good social organisation and behaviour increases material prosperity. When wealth is abundant and widely distributed, its benefits are easily taken for granted. Lush periods encourage the false belief that 'the best things in life are free'. In fact they are the reward for accepting a discipline that can be very costly to follow. Human memories are so weak, outlooks so short-sighted that evidence lying not far away in space and time is ignored.

Moreover Christian concern for the poor and needy can aggravate this trend. For it encourages many to demand ample provision as a 'natural right', 'without strings'. So the welfare policies inspired by Christian ideals and leadership can obscure the source from which they flow.

Dilution

There is an intrinsic difficulty in devising suitable modes of operation for maintaining the quality of a movement when success transforms it from a pioneering activity of a select minority to one adopted by the whole of a society. It then includes large numbers — the majority — who must discharge more static, less exciting roles, without discouraging the enthusiasts. Such a change of state must cause a lot of confusion and tension in any case. This is aggravated when it takes place gradually over centuries, with new situations constantly arising within established groups, new groups joining, and all open to and reacting on each other.

Neither the Church, now that it has become a very large and highly articulated body, nor its individual members should be surprised or disheartened by the occurrence of some blunders and loss of direction at times in these circumstances. Failure to accept the challenge would have been the truly shameful course. Yet neither should they be complacent about even inevitable errors. Alertness to notice, to learn from and to remedy mistakes is just as much their duty as being prepared to take the risks involved in the first place.

Moreover, it is important to note how extremely vulnerable to distortion the comprehensiveness of Christianity makes it. Quite rightly Christian churches have adopted local customs that do not clash with the few essential dogmas of their basically simple creed. They hold that every good thing in life comes from God. So they cannot despise or reject any method of recognising and giving thanks for the benefits received from his hands, and relating the obligations they entail to the patterns and problems of living in many different environments. God in his wisdom has created a wide variety of human beings by allowing them to evolve naturally in different times and climates. If Christians are to expound the boundless love of God adequately, they must make it as readily identifiable as possible to all mankind.

Such open-mindedness is the bounden duty of the Christian Church. Yet clearly it exposes it to a serious risk of admitting incompatible elements engrained in existing local practices. This must be gladly accepted as part of the cross it has to carry in obeying its Lord's command to impart his Gospel to every creature. It must also recognise the dangers inherent in such an outward looking policy, and always be on guard against losing or destroying vital features in the process.

These hazards are greatest at those times when the Church has been most successful. The greater the spread of activity and contacts, the greater must be the opportunities for infections to creep in. That acceptance and expansion must increase the risk of failure and decay is a consequence of its own nature. The Church of Jesus cannot be a self-centred, closed shop. Its loving concern must go out and try to bring into its care those who may corrupt it.

As success compounds the dangers of comprehensiveness, confusion about the real essence of the creed is bound to occur. Many roles in society will be held by weak or nominal Christians; so they will be discharged in ways that bring no credit to the name. Meanwhile Christian officials and organisations will be tempted to undertake tasks in the service of the community for which they are not equipped. These can conflict with their true vocation and responsibilities, and so lay the faith they represent open to misrepresentation.

The Church must face these hazards, and still discharge its primary duty to bring all mankind to the foot of the Cross as well prepared as possible to answer Jesus' challenge 'Who do you say I am?'. To do this effectively, all its members and constituent organs must look very hard and critically at their behaviour. This must be constantly reformed to ensure that it is really conducive to helping others to find the way there, and give the right response. This is not to advocate the fashionable fraud of image-building, but to commend the use of the plain, unvarnished reflection in detecting and removing blemishes.

Intertia

Above all, growth in size causes institutional inertia. This again is aggravated when traditions are clothed with a religious sanction. Many of these have no intrinsic Christian significance. They have been adopted to facilitate comprehension in particular circumstances. The function of any custom must be carefully analysed before it is fostered, or reformed.

This consideration bears on methods of communication. As liturgies are designed to maintain the chain of transmission and strengthen the devotion of committed Christians, style and language must respect the characteristics of such groups. On the other hand, evangelistic efforts are aimed at non-Christians. So these must be conducted in the language, and illustrated by the types of experience that are prevalent in any society. In the developed world these are mainly those of scientifically industrialised economies and urban environments. So unless the Church's teaching can be and is related to the situation in these, its leaders may fall into the error usually attributed to generals in secular armies, of fighting today's battles with the drills, ironmongery and tactics appropriate to the last war.

Present situation

In previous centuries, the suffering caused by natural hazards and disasters raised the most pressing questions about a creed that maintains that the observed universe was created and is sustained by a loving God. It is doubtful if these are now major obstacles to acceptance in the advanced societies. These are where traditional religions are under the greatest pressure. Although they still worry the individuals affected and some special groups, the majority seem reasonably indifferent to such incidents. They take the risk of being themselves involved with equanimity. They may pay some voluntary or imposed taxes to reduce the risks, or support collective relief work. Yet these are only examples of sometimes sensible, often minimal insurance cover rather than of deep concern about the issues. Two developments in the natural sciences have contributed to change in the climate of opinion in which Christian apologists now work.

In the first place, those disciplines have abandoned the absolute, deterministic approach of classical mechanics. They recognise that some degree of randomness in individual events is not incompatible with general laws. Chance variations may indeed play a part in purposeful selection, and the progress of complex entities. Consequently the fallacy of equating benevolence with insulating people from all risks is now usually acknowledged in theory by most thinking people. Only in the political arena and other sectors of society dominated by the advertising media do cruder conceptions of charity still retain their grip.

In the second place, the application of science has achieved a high degree of potential mastery over many natural hazards. This development co-exists with massive evidence of man's inability to control himself, and manage his affairs so that he can enjoy the full benefits while avoiding the dangers that this power entails. This has focused attention on the personal and social rather than the physical flaws in creation.

The callousness of man has always caused much more human misery than 'Acts of God'. The evidence for this proposition is now overwhelming. Christians should welcome this confirmation of their Lord's warning that real evil is a product of men's hearts. The natural order is simply a neutral forum that gives them the opportunity of growing in maturity by trying to overcome difficulties and temptations, or of succumbing to them.

The key to humane solutions of both physical and social problems lies in the way humans approach and try to tackle them. So the practical exposition of any creed given by the behaviour of its adherents is better proof of its truth or falsity than its theological treatises. Attitudes are more cogent than arguments; deeds are more reliable witnesses than words. The relevance of the Christian message is too often masked by the conduct of reputed Christians. Too many scientists allow themselves to be put off by this, instead of insisting on testing the prime evidence for themselves. A strange situation in a reputedly scientific age.

Christian failings

The religious may wrap it all up in the blanket of original sin, their name for the inherent defect in man that blights his best intentions, unless his will is enabled by grace from his Saviour to be conformed to that of God. Such technical jargon needs translating into the common tongue. This is not only necessary to enable the natives to understand the 'Good News' that is being transmitted to them. It is essential for ensuring that conditions within the agency of communication, the Church, are conducive to the successful discharge of this duty.

Its effectiveness or failure is not simply an inevitable consequence of immutable sociological processes working in the societies it serves. Internal factors are significant. The reluctance of the lay majority to recognise the need for help and guidance stems partly from the pride and vanity that are major constituents of man's flawed ego. Yet faulty exposition contributes to their blindness. This in turn is caused by the same defects in the Church's own organisations. They must be identified and corrected.

Arrogance

Most startling and serious of the attitudes that obscure the vision of Jesus that Christians should be setting forth is the arrogance that many of them exhibit. They assert that they worship a God whose love for his creation expressed itself in the almost inconceivable condescension of the Incarnation. They regularly read and recite texts that record that his closest disciples recognised the defectiveness of their human insights, and the folly of being dogmatic about anything but the saving grace of their crucified and risen Lord. Yet Christians have claimed, some still do, authority and infallibility on a par with the pretensions of the proudest heathen priests and emperors.

The absurdity of creatures assuming greater prerogatives than their creator aggravates the resistance that such pride naturally arouses. The acquisition and exercise of power 'for the glory of God' sounds altruistic. It ignores the well attested fact that — if the Christian revelation is true — God himself does not glory in such behaviour. He rejected the suggestion that such means might be used to establish his kingdom in Earth as devilish sophistry at the start of his earthly ministry, and maintained this position steadfastly to the bitterest of ends.

The duty of Christians is not to correct God's mistakes, but to follow his example. They cannot dictate how the Holy Spirit should operate, but must watch humbly and patiently in order to recognise, welcome and assimilate sensitively any and every manifestation of his methods of working. Their organisations must conform to their beliefs if they are to be effective, not *vice versa*.

Too many Christians fall into the human error of thinking that commitment to Jesus must take an almost identical form to their own. They confound loyalty to their Saviour with adherence to their own particular association of worshippers. So their churches tend to become, or to become dominated by inbred circles of religious extremists. Yet the more vehemently any group claims to follow Jesus of Nazareth, the more binding is the obligation not to let pride in the achievements of its heroes, or any intellectual or institutional idol, inhibit co-operation with others who serve him in other sectors in the struggle to carry out his charge to carry his Gospel to everyone.

The Christian churches have moved some distance from the bigotry of earlier centuries. They can now fairly claim that some secular religions have taken over as the front-runners in this particular vice. Yet they should look around and note that they still have a long way to go. There are a lot of historical handicaps to jettison, in their administrative systems as well as their theology.

Faithlessness

Moreover the abandonment of the servant's bench for the master's dais led Christian protagonists into taking up positions that in practice denied the faith that they professed to uphold. Attempts to limit the scope of scientific inquiry, and over-ride its discoveries by the mechanical manipulation of texts, made a complete mockery of any pretence to believe in one God in whom creation lives, moves and has its being. If this phrase has any meaning at all, every properly conducted investigation can only reveal some aspect of God's handiwork and methods.

Undoubtedly the human beings involved will make errors of observation and deduction, and hence propound false conclusions. Concern for the authors of such mistakes must put a great strain on the charity of the Church. Contributing to the scientific effort needed to correct errors and stop the spread of false information must tax its resources, particularly at a time when its introversion has left it ill equipped for this task. Yet Christians have few more valuable spheres of witness in present circumstances. They can no more expect to dodge the sweat and agony of being involved in the real world that their Master did.

Faith is not the enemy or converse of reason at all, but an essential component in any rational philosophy. It is not measured by the extent to which one can be persuaded to reject evidence and assent to irrational propositions. On the contrary, it is necessary for sustaining any thinker. It should remind him that the most vital knowledge of all — knowledge of one's own self — might be described more accurately as knowledge of one's own limits and ignorance. Its function is not to contradict insights acquired by observation and reflection. It should provide the discipline under which diverse activities can be held together and given a sense of purpose, constant warning of the constraints under which data has been collected and must be interpreted, and principles to guide the making of decisions in spheres where pure science can offer no criterion.

Such a balanced attitude is not promoted by high priestly claims to authority that is rapidly shown to be worthless. It is encouraged by readiness to admit the difficulty of comprehending, and hence the possibility of misunderstanding many aspects of our Creator's methods in any transient state of partial knowledge. This more sober approach sacrifices no vital position. It prevents the demolition of untenable ones being used to distract attention from the impossibility of accounting for the existence of the universal belief in the existence of moral standards without some metaphysical source from which they can draw substance and authority.

Those clergy who left their service function in pursuit of secular power sold their birthright to the scientists. They must now face the awkward fact that, by the tests their Lord recommended, engineers and other applied scientists are today exhibiting a livelier faith than they do. It is these professions that are moving mountains, and providing the means of healing the sick and feeding the hungry. Their proceedings are often the only ones that penetrate barriers of politics, race and language freely, and so promote genuine respect and fellowship between all sorts and conditions of men.

Perhaps a more radical reform of theological colleges (if such divisive institutions are to be retained at all) than is likely to stem from co-operation between denominations might be achieved by attaching them to technological universities, instead of traditionally oriented ones. A full measure of practical experience is also desirable before allowing candidates for 'holy' orders to pontificate on matters outside their field of competence.

It is indeed strange how many Churches, even among those that formally admit the primacy of the Bible, ignore the evidence it presents of God's nature, of how he thinks and acts. He himself spent most of his earthly ministry as a simple village carpenter, framing buildings, and making and mending domestic furniture and tools. To carry on his work he chose men who brought considerable, varied, practical experience to the task. He did NOT collect a group of enthusiastic young men whom he could easily mould in his image.

Constantly recurring indifference to God's own example, and preference for models developed by secular agencies, may explain the difference between the impact he made on a heathen world with apparently quite inadequate resources, and the inability of highly equipped Churches to hold the interest and loyalty of nations already conditioned by Christian ideas.

The belief in an orderly revelation of an ordered creation postulates a greater miracle than the kind of arbitrary interference with the natural order that is popularly credited with the title. Charles Darwin standing spellbound in wonder before the beauty of a single flower followed faithfully the example of Jesus. So he was rewarded with deeper penetration into the mystery of how God's nature works, than those who tried to curb his freedom of inquiry by legalistic argument and sophistry.

The Christian churches are committed by the precedents set and the precepts expounded in the life and teaching of their Lord to using their eyes and hands in his service. Used with humility and dedication they can provide otherwise unattainable insights. There is much still to be done to erase memories of earlier, and some continuing, attempts to deny the facts of life in defence of vested interests.

Lack of charity

The spread of education and the work of historians and translators is making reference to the most trustworthy records of the ideas of Jesus, the Christ they claim to follow, readily available to everyone. So the uncharitableness with which Christian churches have refused, to some extent still do, to work and worship together destroys much of the value of their witness.

Differences of outlook and organisation are neither surprising nor sinful. Any living movement designed to embrace every type and condition of human being must produce a complex pattern of ideas and associations. Institutions must accept that the Creator did not make all men and women equal and of the same type. He simply invited all into the same loving relationship with himself, while leaving their personal characteristics all individually, widely and delightfully different.

Differences of vision and practice are to be expected, indeed welcomed. For the Church militant here in Earth needs a large number of specialist and an enormous number of non-specialist units in its ranks.

The scandal of present divisions does not lie in the variety of organisations, but in the aggressiveness with which claims for pre-eminence are pursued. Even worse, in view of every sections' protestation of its unworthiness and utter dependence on the grace and mercy of God, is their refusal to join one another at their Father's table.

The warnings of church authorities against the dangers of uninhibited fraternization may conform to organisational theory, be good tactics or shrewd business. Indeed they can be defended by many conventions commonly adopted by the flawed human race. They are quite incompatible with the loving concern for all mankind that Jesus insisted was the primary mark of being one of his followers.

As the words and actions used in corporate worship should reflect the convictions of the celebrants, Christians might well consider whether confirmation of the sincerity of their acceptance of one another as brothers and sisters by a period of open communion would be a useful preliminary move before greater organic unity is attempted. As well as testing themselves, this would demonstrate to others that the proposals are motivated by a real spirit of charity, not just a manifestation of the fashionable trend towards bigness, of amalgamations and takeover bids aimed at greater power for in-groups of ambitious men.

Critical appraisal

The best evidence, the earliest records of the practice of the Church when it was most directly under the influence of its Lord's physical presence on the Earth, shows that these repellent features are not truly elements of the Christian faith at all. They are relics of phases when churchmen slipped into secular roles and attitudes. These were at best only marginally or transiently compatible with their vocation. They were often completely at variance with their claim to be members, let alone examplars, of the Church of Christ.

During a long period clerics were the only educated sector of European society. They were struggling to graft their message securely into the conscience of tribes that were still savage and heathen at heart. So for understandable and occasionally commendable reasons, the Church adopted many dubious customs and extra-religious responsibilities.

At that time these may have helped to guide men to the Christian faith, to stabilise their loyalty to it, to safeguard its teaching and the development of social behaviour animated by it, or to discharge what seemed inescapable duties to the community. Many are now an obstructive, irrelevant overburden hiding the truth that the Church should be exposing. Totally unchristian attitudes are often generated from secular precedents that have been included.

Institutional inertia and sectarian interests have combined for centuries to proliferate defences of untenable positions and blown bridges. Not only between Christians of different styles and outlooks, but above all between most of them and unchurched or marginally attached sinners. God, in creating humans in his own image, created most of them of this type. Jesus specifically claimed that his ministry was primarily directed to them.

Proposals to jettison encumbrances to co-operation in the work of bringing the knowledge of Christ crucified to those isolated from him by ignorance or indifference clearly embarrass the Churches today as much as this radical approach by St. Paul upset some of his colleagues in the Apostolic era. But the early Church had the courage and sense to ride the shock his initiative caused, stick to the tests their Lord had given them for sorting out what mattered, and so set in motion the first outburst of missionary activity from its Judean home.

It was one of the main wings of this movement that eventually managed to discipline the turbulent inquisitiveness of the European tribes sufficiently for them to generate, when more mature and given the right conditions, modern science. Their successors in title have a great responsibility for recognising that the outlook of those first pioneers is still essential if many of the objections to the faith raised by those following the system it fathered are not to be sustained; and if they themselves are to make their due contribution to understanding and helping to control the tide of events that it set flowing. As they are very firmly reminded, vines need regular, drastic pruning of mere foliage if they are to continue to bear ample, good fruit.

Secular failings

All the faults do not lie on clerical shoulders. The scientific community shows a lack of historical perspective that does little credit to our educational system. This prevents many laymen understanding processes that operate in an evolutionary manner. It has allowed many to accept, openly or unconsciously, the Marxist dogma that all religions other than itself are simply opiates brewed up to control the mass of the populace. Such emotional, unscientific reactions to past mistakes obscure the lessons that must be learnt from them if even greater ones are not to be made in the future.

In the initial stages of any new line of inquiry and development, some glimpses of the truth are all that can be expected. At the take-off stage, the co-operation of many with only a minimal attachment to the ideas tested, of many only jumping on the band-wagon for fun or profit without any concern for deeper issues involved, is often unavoidable. Indeed it is usually essential if the progress needed for fuller comprehension is to be made. Such allies can precipitate awkward situations, and for a time lead people down false trails.

Yet such consequences of inability to see immediately the full, completed picture, whether in economics, engineering or theology, are not evidence of the lack of any valid elements in the theories being propounded, even less of the insincerity of their proponents. Some degree of commitment to partial insights is usually necessary for making the progress on which deeper penetration depends. Some errors due to inexperience and lack of knowledge are inevitable during the pioneering phase. The right antidote to the ills these cause is alertness to notice the symptoms that expose the limitations of elementary practice, and readiness to correct dangerous trends before they get a grip and do irreparable harm.

In every sphere of human activity, eternal vigilance is not only necessary for real progress. Without it even the advances that have been made will be lost. Responsibility for many of the blots on our landscape lies less heavily on our forebears' sometimes too ready acceptance of first approximations as final solutions, than on our vain delusion that our way of life is a natural birthright, requiring no sacrifice to maintain it. We have not so much rejected the faith of our fathers as let it slip through our feeble fingers, by falsely assuming that its benefits could be retained without bothering how and why.

Current opporunities

Yet science functions by continually asking just such questions. So a scientific culture should give the Church the finest opportunity for rekindling concern about fundamental issues since the conjunction of Roman order, Greek thought, and Hebrew spiritual insight formed the carefully prepared seedbed for its germination and initial growth. But it must be properly equipped and trained, with decks cleared for action, if it is to seize it.

The uselessness and inconvenience of the material standards on which most current theorising in the social 'sciences' is based are now obvious. So there is widespread awareness of the need to reform attitudes to life and its environment, if our mechanical triumphs are to improve personal freedom and personal relationships rather than to threaten them. That some earnest seekers of the way to achieve this turn away from Jesus is not surprising, when his disciples have allowed his body in Earth to become so badly infested with parasitic growths.

The Church Militant needs a John Moore to scrap obsolete paraphernalia and tactics, and scrape off the barnacles that have accumulated during the centuries in which it has lived on the capital won by past efforts. It must become a mobile, efficient fighting force if it is to keep on top of the tricks of its old enemy. In such a context, many if its critics might — like Saul of Tarsus — come to realise that they had been trying to advance in the wrong direction, and similarly turn about to take a leading part in keeping the Church moving forward on the right lines.

Christianity is the system of belief that most effectively holds together in a coherent relationship all those principles that are essential to modern scientific thought and practice. It provides rational grounds for the beliefs about the nature of the universe and the efficacy of scientific methods. It gives authority to the codes of personal integrity and fellowship that are necessary for its successful pursuit. It offers some hope that all the apparently conflicting claims of different groups in which one recognises some element of value will ultimately be reconciled. To accept long term results and discount many transient errors when preferring medicine to witch-doctoring, or natural science to alchemy, while rejecting equally cogent evidence for the validity of Christian experience is schizophrenic rather than scientific.

Science can never prove any religion true or false. Yet it can and does expose gaps in its own foundations that cannot be made good with its own equipment and techniques. As scientific progress depends on freedom of inquiry and the development of mankind's talents to the utmost, all members of a civilisation that hopes to benefit from the application of science to its affairs should be concerned to maintain the authority that supports such liberties to the highest degree compatible with respect for other people, and which stimulates the social co-operation that makes human existence possible at any but the lowest subsistence levels. Absolute freedom is as unattainable in practice as other absolute ideals once sought, but now recognised to be beyond the grasp of mortals. It is stupid to discard real, positive advantages in pursuit of a will-o-the-wisp.

Too many scientists confuse dogmatism and sectarianism with religion. This is an interesting example of the non-transferability of mental skills between different fields of inquiry, as weeds of the same genus flourish in their own cabbage patch. They are right to reject distortions; but they must take care not to cut off the branch on which they sit, by trying to isolate themselves from the main stream of human experience that justifies and supports their local bit of the system.

Modern science is Christian science. If it does not respect its parentage it will inevitably become an uncontrollable menace that no account of ingenuity will be able to direct for the use and convenience of mankind.

Chapter 11

No man's land

The provision of factual evidence to support otherwise abstract argument is normal practice in the applied sciences. Here it is drawn from 'the English disease'. This is the outstanding current example of communal schizophrenia, of collapse into anarchy caused by allowing links essential to social stability to corrode away or be destroyed.

This is not surprising. The United Kingdom has never been a great power in physical terms. The economic strength that allowed it to assume such a role during the 19th century sprang from its pioneering the Industrial Revolution. So it could have been expected to hit problems that flow from careless over-industrialisation earlier than other nations.

In fact, some of its more energetic competitors overtook it. So in this century, Britain has twice had to take a leading part in stopping the menace of unbridled mechanical power. The effort left it weak, and susceptible to the contagion. So the affliction now carries its name; England is now the best field for research into its cause.

As in the preceding chapter, some religious symptoms will be noted first, afterwards secular ones.

The English Church

England emerged from the upheavals of the Reformation/Renaissance in a much happier state than most of her neighbours, thanks to the combination of its situation, its temperament and the genius of the Tudors. Persecution and bloodshed had been on a much smaller scale. So the nation was neither torn apart nor deprived of vital elements. Although sects of every variety, native and imported, ancient or modern, flourished in its fertile soil, *de facto* tolerance was practised well in advance of *de jure*, provided heresy was not compounded with treason. This allowed many who might have been alienated or driven into exile to remain, make their fortunes, and contribute to the life of the community.

At the heart of this development, the national Church — freed from its foreign entanglements — was able to revert to the more easy going, comprehensive style it had lost at the Norman conquest. It provided a vehicle capable of sustaining the faith of many enthusiastic non-extremists, as well as the vast majority of non-enthusiasts.

Four reforms contributed to the process of reconciliation. These had been grafted on to a parochial system intimately related to the social structure in units of humanly manageable size. This is a necessary condition, if ministers are to be able to know their flocks and truly care for their souls.

1. The open Bible, superbly translated into the common tongue, ensured that the best record of the Christian revelation of God was readily available to everyone.

2. A system of common prayer, with a memorable liturgy based on the best of the traditional formularies and reformed in the light of the Scriptures, provided both a personal mirror and a national focus for religious life. The former is essential if the lessons learnt and the convictions expressed on Sundays in church are to be reflected in the personalities of Christians, and in their relationships with others in the bustle of normal, everyday activities. The latter is necessary if any social unit is to maintain a Christian commitment and ethos.

3. A ministry educated alongside and closely related to the community in which it worked enabled clergy and laity to understand one another and work intelligently together. The new grammar schools and the ancient universities served both groups. Practical training on the job discouraged the development of affectation and pedantry. Sharing work and pastimes helps to strengthen affection and deepen insight, as a carpenter in Nazareth, a tent-maker from Tarsus, some fisherman on the Sea of Galilee, even a tax collector discovered centuries ago.

4. An Erastian system of government provided an excellent device for curbing clerical excesses, an essential one for national recognition.

These features enabled the Church of England to play the central role that only such a national institution could in instilling Christian ideas and attitudes sufficiently generally throughout the nation for them to provide sound intellectual and social foundations for the application of scientific methods to production and political problems. They facilitated Britain's early achievements in the development of industrial techniques, and in the institution of social reforms peacefully; but success led to the growth of complacency and official arrogance.

The sloth and snobbery resulting from dominance allowed the Church of England at first to ignore the growing industrial population. Then, despite a few prophetic voices, to abandon it to nonconformist ministries. Finally it retreated into neo-primitive ritualism and legalistic, pre-scientific dogma, when it realised that it had missed the bus. The story can be traced in the history books.

What is relevant to consideration of the possibility that the faith is necessary for the safe operation of a technologically powered society, and if so how it might be re-activated to work effectively, is to note the vital elements that have been discarded. For control systems cannot function if their guts and communications are ripped out.

Ministry

The most fundamental of the mistakes was the reversal of the policy for training the ministry (3), by the creation of theological seminaries. This disaster was aggravated by the fact that many of them were founded with a specific sectarian bias. Such institutions inevitably tend to develop those formal aspects that are of interest to full-time professionals, particularly when they are located in isolated monastic situations. Yet these are of little interest to the laity. So emphasis on them causes a fatal separation between elements that must work together.

As with the Janissaries, such a system of recruitment and training may be an excellent way of producing a corps dedicated to the protection of the powers-that-be from the people. It is useless as a means of inducing understanding, and the art of leading by service.

The divisive effect of this departure was greatly aggravated by the respect that the humble, in Chaucer's day, parish priest had acquired in becoming 'the parson'. For this encouraged some of the colleges to imagine that they were training a corps of officers. In commending this view of the ministry as commanding respect and attracting men of high quality, bishops forgot that this was not the quality of Christian ministers, if the Christ they followed was Jesus of Nazareth. When this rejection of Our Lord's clear example and teaching was piled on the traditional English distrust of standing armies, it was not surprising that a priesthood in which a high proportion were seduced by such foreign ideas alienated all but the small minority of the natives addicted to self-flagellation.

Of particular concern to the present time and thesis, ministries cannot give pastoral guidance to their flocks if they cannot understand the language in which many problems that arise in an advanced economy must be framed.

Once, not very long ago, it was one of the glories of the Church of England that it included a strong contingent within its ranks who were mathematically fluent, and so could provide enlightened guidance. Those who enjoyed personal contact with such teachers know how deeply they enriched the educational system. The lack of similar leaven today is regrettable. The tradition should be restored. The retreat into a purely literary, theological laager must be reversed — unless the Church of England is determined to commit suicide.

Worship

Moreover the arrogance that the officer class mentality quite naturally engenders played — and is still playing — a major part in the destruction of other features that once enabled the Church of England to minister effectively to the English, and so hold their allegiance to the Christian faith.

In their conviction that they are the experts, and entitled to issue orders, the clergy have gone a long way towards ruining the system of common prayer (2). Rejecting 'Thy will be done' in favour of the fashionable 'I did it my way', many have abandoned one of the great liturgies of the Christian Church, despite the assurance of their Bishops that this would not be done when the delegated legislation under which it has been done was allowed to stand. Its place has been taken in many areas by a rather mundane Do-It-Yourself manual that rivals the infamous 'Rules called the Pie', in making it 'more business to find out what should be read, than to read it when it was found out'.

The decline in the ability of the English to read, write and speak their common tongue with any degree of competence to which this contributes is indeed more directly the responsibility of parents and schools. Yet there seems no good reason why the Church in its charity should abandon entirely its traditional interest in the provision of sound standards of general education. The trend exposes two ways in which the segregated education of clergy prevents them understanding those for whose souls they are charged to care; so that the flocks they are called to feed and tend scatter far out of their reach.

As St. Paul pointed out, while any form of exuberance and

language is legitimate in private prayer or carefully selected groups, public worship should be conducted decently and in order. This condition cannot be met when the rate of change and the multipicity of variations, sometimes confusingly similar, sometimes so strange as to leave many bemused, often cause experienced ministers to stumble along in an unedifying way, losing their congregations in the process.

Moreover most of those who prefer the ministrations of the national Church to any of the wide range of alternatives on offer expect their loyalty to it to be matched by its to itself, and them. The English do not like being made to look foolish on public occasions and on their own home territory, as happens when they have to try to adjust to varieties from A to Z, to Bishop P, Canon Q, or the Reverend Mister R when they have to move or suffer a change of incumbent. Those who drop out under such pressures may be the weaker brethren. Yet the Scriptures make it clear that it was to save such sinners that Jesus lived, died and rose, not to found a championship team of spiritual athletes.

The chaos is often justified by the claim that the language of the Book of Common Prayer is no longer understood, its traditions no longer valued. One benefit of a stint helping to transform a large technical college into a modern technological university was an opportunity for testing such assertions in a more scientific, less emotional atmosphere, and a more democratic, less hierarchically bound constitution than the Church now offers.

Similar arguments were advanced as loudly there as in the Church. Despite using much less digestible test pieces (Bacon instead of Cranmer) on notoriously illiterate material, the falsity of the incomprehensibility argument was immediately, almost embarrassingly apparent. This result should have been anticipated by anyone familiar with the awkward jargons that the young very quickly master if they are convinced that entry into the mysteries they encode is worth their while. The falsity of the argument against traditional practices was revealed by the ease and absence of dissent with which students who were reputed to reject antiquarianism totally took to conventions, styles and even vestments much less compatible with the 20th century than the Book of Common Prayer.

Even more damaging than the disorder, disrespect for valued traditions, and debasement of language, has been the trivialisation of the sacrament by which Christians proclaim their Lord's death until he comes, and share together in his risen life.

The Reformers, aware from direct, personal experience of how readily familiarity breeds contempt, and mechanical gestures can turn into idolatory, emphasised the vital importance of preparation, and a spiritual atmosphere in which communicants can discern the sacrificed body of their Lord, and so eat the bread and drink the cup worthily.

A heedless momentum has been allowed to distort the intention of saints like Keble, to restore the Eucharist to its proper central role, by confusing frequency with quality, mechanics of ritual with reality of worship. Now a 'chips with everything' approach, and the hustle involved in trying to fit everyone into the scenario, disturb and distract those who would join the faithful handful at the foot of the Cross, as surely as the mob did on the first occasion. And the perpetrators — like those of the specification so comprehensive that the plane could do everything but fly — are then surprised when all but a small minority leave them to tithe their mint, dill and cummin behind invisible walls.

Structure

Adding insult to injury, the extent to which the changes have been foisted on the laity, many of whom respond by voting with their feet, is obscured by the pretence that the erosion of parliamentary control (4) over the national Church has been achieved by democratic processes. In fact the great majority of members of the Church of England are completely disenfranchised. Not even those on electoral rolls have any vote in the elections for the House of Laity of the General Synod, the body that is claimed to reflect their views. The complex, poorly advertised system of indirect elections, and the massive blocking powers reserved for self-perpetuating groups of permanent officials exceed anything that even the most brazen trade union bosses have ever dared to propose. The growing bureaucracy by which — almost unknown to the people regularly in the pew, let alone the more casual attender — a mass of legislation that is rapidly changing the very nature of the Church is processed, is dominated by similarly unrepresentative cliques.

The most serious consequence of the amount of time and energy now devoted by the shepherds to demonstrating their professional expertise and playing ecclesiastical politics has been the reduction in that available for reading the Scriptures, setting them forth by their lives and doctrine, and counselling their flocks in their light. Despite the availability of a range of translations that makes the Bible accessible to anyone who can read, the lack of encouragement to work constantly and thoroughly through it again and again, and of the exposition and discussion needed to complement and deepen private study leaves many ill-equipped — individually and corporately — to apply its examples and teaching to current problems.

Yet if the belief that events which occurred nearly 2,000 years ago formed a climacteric in the evolutionary historical process of eternal significance is to inspire men and women, they must be enabled to see both the continuity through the ages and the potential for the future that the Spirit offers for leading them, slowly but steadily into all truth.

The Church of England has by now slipped back into most of the errors that came near to destroying the hold of the Christian faith on the English *et al.* in the 16th Century. As it no longer enjoys either the feudal authority to enforce or the psychological power to frighten people into obedience, it is retreating with what it has managed to salvage of its inherited wealth into a pietistic and sectarian ghetto. Such sanctuaries are seldom comfortable or secure. It would do better to turn again. It cannot discharge the trust placed on it by its Lord unless it does turn again. Achieving this manoeuvre successfully will require a strong, prolonged, joint effort by enlightened clergy and laity. On the former occassion it was those open to new learning who took a leading part. So in the 20th century the task will need a major contribution from the most advanced sectors, which lie in the applied sciences.

English Education

Education is now essentially a secular function in England. A few pockets of religious influence remain; but even these are dominated by bureaucratic and financial controls devoid of any religious, or even moral commitment.

Two examples, one from natural, one from social science, show how far short of instilling comprehension the system falls. Yet such understanding is essential if any community is to be able to apply scientific knowledge with sensitivity and intelligence.

Most presentations of the voyage of the Beagle suggest that the theory of evolution to which it led conflicts with Darwin's hope of proving the truth of the Bible . Yet this is just what it did, once the fog of clerical obscurantism had cleared. That book is the greatest evolutionary treatise of all time. It demonstrates how careful selection over millenia enabled mankind to develop the highest and most elusive of all human faculties, that knowledge of oneself and of one's relationship with one's Creator that is rightly entitled 'wisdom'. Like some of the great cathedrals that have had to be underpinned and reinforced to enable them to survive despite their medieval builders' ignorance of soil mechanics and structural engineering, it now rests on firmer foundations that Archbishop Ussher's valiant exercise.

Similar distortion of language has tended to detach from the epithet 'human', with all its emotional charge of love and hope, any idea that behaving intelligently and responsibly is included. So it is now commonly implied that to be petty and irrational, to lack foresight and integrity, are endearing human characteristics, while anyone advocating or practising the converse virtues is an inhuman robot. Thus one of the ablest and most honest post-war leaders of a political party could be dismissed as 'a desiccated calculating machine', for suggesting that his party ought not to fudge its arithmetic, but set out its prospectus as frankly as it would demand that other sectors of the community should.

Were the second, unflattering misrepresentation valid, there would be no rational grounds for believing that the human race can survive. That both persist indicates the need for urgent reforms. For unless the emergent generation is provided with a sound base and adequate equipment, it cannot make a useful contribution to the evolution of society. Those ill prepared will lack the personal motivation needed for full development. Those with an unbalanced approach may well run amok, eventually destroying instead of helping to rebuild.

Proper educational provision is vital in societies that aim or claim to adhere to democratic principles. For while the problem might be containable in a Platonic autocracy, as no one incompetent mathematically would be admitted to the corridors of power, it fatally undermines democratic systems. Government of *all* the people, by *all* the people, for *all* the people (the original version, that Lincoln felt obliged to omit the 'alls' when dedicating a civil war cemetary is understandable) of a scientifically sustained economy is manifestly impossible inless *all* the people have sufficient understanding of the language in which key problems must be posed to be able to assess alternative solutions intelligently. Denying *all* the people access to mathematics not only freezes the genial current of the democratic soul by withholding knowledge. It prevents them realising how easily the deficiency could be rectified, for innumeracy is no more a genetic defect than is illiteracy.

The inability of the United Kingdom to develop a reasonably united response to its present opportunities exposes three major deficiencies in its educational system. It is not enabling children to respect the pattern of behaviour on which its own past successes and the scientific system alike depend. It is not enabling them to acquire the knowledge needed to understand their situation. It is not enabling them to learn the language of the technology on which they rely. So it is not surprising that large sectors are falling into the derelict, disputed wilderness of a no man's land.

Behaviour

A comprehensive system of education is clearly needed. Yet a comprehensive system and a system of comprehensive schools are completely different, incompatible ideals. Indeed the latter is a figment of the imagination, as unattainable as the universal solvent or machine tool. The pursuit of the phantom merely wastes resources urgently needed in the main task. Even worse, it prevents schools being designed and deployed as co-operating units in a truly comprehensive system.

For the key to the scientific reformation that has precipitated the present flood of information was not itself the finding of some formerly secret formula, or source. It was the adoption of a particular pattern of behaviour. The conventions observed are still essential, not only for maintaining and managing the flow, but for choosing which items from the existing stock are appropriate for handling any specific problem.

Patterns of behaviour are initially acquired by example and experience. Analysis and development must be built on these, so only become feasible later, much later. To enable pupils to do this, the school system must provide units of a size and ethos that helps children to come to appreciate how and why institutions work. They must also be allowed to stay long enough to absorb the lessons of one stage before embarking on the next.

Children do not acquire the art of discrimination by being dumped in an information super-market before they can even read the gaudy labels, let alone have any idea of the value of the contents.

Human understanding

In the acquisition of knowledge it is important to restore the old 'matric' (c.f. the foreign 'bac') ideal, of insisting on a grounding in a broad enough range of disciplines at the compulsory first stage of any schooling. Before specialisation is allowed, all children should be encouraged to acquire, those going on to higher education should be required to acquire sufficient basic knowledge of sufficient facets of their common culture to make communication between different groups feasible. There are six main spheres of human experience with which adult citizens need a minimal familiarity if they are to be able to understand their environment, and fit their own careers and interests — even the changes they wish to introduce — into it effectively. A hexagonal common core is essential:

1. The native language
2. Mathematics

These are the languages in which most of the information people need is recorded. So ability to read and understand them, and some degreee of competence in putting ideas into them is essential. Yet teaching each of them suffers from two major handicaps:

The first is that both are now so highly developed that the basic sectors are no longer of much interest to those who have specialised in these fields. Such guides can be tempted to neglect important aspects in favour of less relevant novelties. Hence the need to feed-back a higher proportion of those who have graduated (and have had experience) in the applied fields into the schools.

The second is the debilitating influence of electronic devices, the goggle box and the calculator. Immense efforts are needed to counter commercial pressures, to ensure that students continue to develop their own mental abilities and range of reading, so that they remain masters of their mechanical aids, not slaves.

3. History
4. Geography

The former is necessary to enable students to acquire some feeling for time-scales and how human relationships develop, and in particular to recognise the threads that are woven into their own backgrounds. The latter develops a sense of spatial relationships, and the physical constraints within which people must live, if they are to live peacefully.

5. A natural science, to introduce all to its methods of analysis, of measurement, and of testing theses.
6. A foreign language, to counter insularity

As experience has shown that most pupils capable of benefiting from higher education can manage 7 to 10 O level subjects, insisting on this basic half-dozen imposes no hardship. It still leaves ample scope for adding another language or science, and non-academic subjects such as religious studies, arts or crafts. The range of options at this level would indeed be reduced; but the present number of options available has no merit. Many offered and taken have neither rationale nor relevance to the pupil's needs. They confuse parents and pupils, and distract attention from the importance of planning their education and lives with some degree of coherence and foresight.

For those with less academic bent, courses covering the same ground but to less depth should be provided, with their own related examination system. The suggestion that the whole range of ability can and should be included in a single system is another of the universal panacea fallacies. If some fail to cover the central hexagon completely by the time they leave school, they must not be classed as 'failures'. They will have had a sight of prospects that may materialise one day. In the meantime they share a common background with all their fellow citizens, some vision of their joint heritage.

Many pressures, such as the growth of bureaucracy and the populist drive to debase standards in pursuit of an unattainable equality, discount the real quality of the young too readily. If teachers are to draw out their maximum potential, they must not be afraid to aim high. Not reaching all objectives by a fixed time is not a disgrace. All must progress at their own individual speeds. The effort devoted to trying is NOT wasted; it can bring its own rewards in due course, and inspire further endeavour. Accepting decline and defeat too easily, and settling for a comfortable ride downhill, is one form of the English disease that we have tolerated too long.

Understanding technology

One feature of technology that makes it particularly difficult to control is that its ramifications extend so widely. Often the amount of information involved in any problem can only be recorded and handled conveniently in mathematical language. Moreover, the more it develops, the greater the weight and complexity of the data necessary for adequate presentation and analysis become. So unless the ability of the governing strata of any society to understand the *lingua franca* of modern science develops at a corresponding rate, informed, rational, sensitive control of technological systems becomes impossible.

The misinterpretation of the significance of the traces of economic growth discussed in the final chapter is a classical example of the danger of a discrepancy between the possession of information and the ability to interpret it accurately. Like the Delphic prophecies, some statistics have been taken as the green light for unrealistic, euphoric expectations, instead of as warnings. So the rape of the Earth's natural resources is still pursued with a short-sighted intemperance that has already left the earliest, European exponents disoriented and vulnerable, and once nearly reduced the magnificent heartland of North America to a desert.

The United Kingdom has squandered many of its raw materials and much of the wealth earned by its industrial pioneering on building and losing an empire. It created an Establishment for this purpose trained on antique models rather than modern techniques. So it now finds itself trying to cope with a phase in which considerations of scale make some governmental interventions in the economy essential with innumerate governing classes. This is a situation that reduces its management of technological processes to a series of reckless gambles.

A grim, closer example of the crippling effect that lack of this vital element in the body politic can cause is inflation. Only communities without even elementary comprehension of mathematics could have remained so long indifferent to the constitutional sclerosis it causes.

The absence of any reliable standards of value makes it quite impossible to compare alternative technological proposals rationally. Yet this is essential if the probability of any positive contribution from each to the welfare of the community is to be determined, and the most promising investment of resources chosen. The deficiency paralyses the central nervous and control systems of the economy, and leads to the waste of human and material capital on schemes that intelligent appraisal would have saved.

The greater complexity and subjectivity of the concept of social value makes definition and refinement more difficult than (say) standardising the length of the foot. Yet the abandonment of primitive bases like gold without providing any better, agreed, constant alternative medium of exchange makes economic assessment and honest planning impossible. Just as modern engineering would be, if works had to be set out with an unrelated multiplicity of tatty, perishable, elastic bits of string, whimsically recalibrated at random intervals.

Morally the corruption inspired by the toleration, indeed the promotion by some public figures, of institutionalised bilking through debasement of the currency, has been a major factor in the decline in decent, responsible standards of conduct. For public and private attitudes do interact. Within the general decay, two examples of current lack of respect for the honest use of symbols are particularly striking:

The first is the claim by advocates of inflation to be 'Keynesians'. Keynes himself described attempts by better organised sections of the community to benefit themselves at the expense of others as 'group selfishness hustling someone else out of the queue', and 'a mug's game'. As one thing he was NOT was a mug, players of that game have as little right to claim to be his followers as Torquemada had to be a disciple of Jesus of Nazareth.

The second is the claim of some of the fat sheep as they elbow their weaker brethren out of the trough to belong to 'caring' organisations. Any group that uses its muscle or its blackmailing power to get a rise in pay higher that the rate at which the national wealth is rising is in truth announcing that it intends to grab more regardless of who gets pushed to the wall.

Much more must be done to open the eyes and sharpen the understanding of *all* the electorate, so that they can see what is happening, and cannot be 'blinded with science'. Experience of several levels of education and their products suggests that three reasons of the present high level of innumeracy need more positive attention, while one fashionable innovation needs more careful control.

Undervalued:

Firstly, all must be given a sounder grounding in handling the numerical and spatial symbols and relationships that constitute the basic vocabulary and grammar of mathematics. Even if, as with literary languages, few reach any high level of ability to articulate it skilfully themselves, they should be encouraged to persevere as far as possible, and must be helped to reach a reasonable standard of ability to understand it. As with all practical arts, this will involve discipline and practice, to which some will take readily, some only reluctantly. Yet leading all kinds of pupil through whatever initial barriers inhibit their progress to a stage whence they can educate themselves is the true function of the educational system, not indoctrination or simply keeping them amused and off the streets. Those who are not led safely through the wilderness into sight of the Promised Land will remain mentally retarded, mere operators of processes set for them by others, and easily deceivable by every huckster who crosses their path.

Secondly, in order to enable all to gain some appreciation of the limits of reliability of much public information, and of the indeterminacy that exists in the universe, some acquaintance with statistics and probability — with 'fuzzy numbers' — is essential. In this field mastery of techniques is not of prime importance. Indeed too much confidence in accepting or rejecting evidence on the strength of elementary, easily teachable standards can be a menace. What is needed is exposure to a wide range of examples, to bring home the need for careful discrimination in choosing between the very different standards of judgement appropriate to situations of widely differing frequency and importance.

Thirdly, as all are bound to be exposed to packages of ideas to which handy labels have been attached for easy use and saving time, all must be taught to ferret or work out the limits within which these can be used safely, so that exaggeration and misapplications in sales promotion can be quickly detected. The hyperbolical character of Archimedes' lever is readily apparent, so that it can stimulate comprehension of the power of levers without anyone forgetting that the provision of a suitable fulcrum, and of adequate size and strength for the lever would cause difficulties in practice. The limits within which many mathematical ideas, statistical techniques and computer programmes are valid are neither as obvious nor as carefully publicised.

One of the most intriguing examples of this last problem is the popularity in some clerical circles of the term 'lowest common factor'. It is doubtful if the users of this pseudo-mathematical phrase realise that it is nonsense, except in the very specialised sense of the unit operator. That might be equated to St. Paul's 'Nothing but Christ crucified'. Yet it is usually trotted out by defenders of sectarian paraphernalia to stir up emotional opposition to his 'highest common factor' approach to spreading the Gospel.

Oversold:

On the other hand, more care is needed in using visual displays and calculating machines, particularly of the 'black box' type. For commerical pressure and anxiety to be 'with it' are threatening the development of qualities that are essential if more children, eventually all children, are to become numerate.

One of the arguments advanced to justify early, general use of such devices in schools is that they are now universally available and used in working life. So in the belief that their primary function is to prepare children for adult employment, similar modes of operation are adopted in learning as in practice. Which is odd in this subject. A careful, logical approach to problems is claimed to be one of the benefits of studying mathematics. So these qualities should be evident in its own arrangements. The argument is as faulty as suggesting that because in later life pupils will probably do most of their travelling by mechanical means, they do not need to learn to walk, run and swim properly, and keep both these skills and their limbs in good fettle by regular exercise.

Laying firm foundations in any field requires different techniques from those appropriate to more advanced stages. The acquisition of essential attitudes and basic skills depends on self-discipline and a sense of responsibility. These are picked up mainly from the ethos and standards of behaviour of the institutions attended. They are also promoted by a traditional, practically oriented curriculum, provided teachers have the experience needed to draw out the relevance of different facets of different situations, and the importance of developing speed with accuracy in mental operations by constant practice. Among the latter, fluency in English and mathematics are the most important. They form the bedrock on which professional communication and co-operation must be based.

Understanding of and familiarity with shapes and symbols, and with the operations that can be carried out with and on them, can only be acquired by working constantly with them, not by pushing buttons that churn out answers like rabbits from a magician's hat. So even aids such as slide-rules and tables, which develop students' insight more effectively by displaying quantities and relationships visually and making them work out their own programmes, need careful introduction and control if they are to reinforce mental growth, and not allow vital faculties to atrophy for lack of use.

While it is sensible to introduce children to modern devices, and give them some idea of how and when different types can be profitably employed, extensive use and reliance on them must be avoided until the students' own mental abilities have been sufficiently developed for them to understand what they are doing and why, when they adopt any routine. With most children, this stage is not reached until late in their school careers. Until then, if numeracy is to be improved, mental effort must be demanded, black boxes remain in the background.

In particular, care must be taken over the admission of such devices to examinations. For the object, the sole justification, of such ordeals is to test the students' progress in understanding the subject and in performing the operations involved. This cannot be done if their personal logic, knowledge and operational skill is over-ridden by the programmes and mechanical reliability built into a calculating machine. So in most school examinations, mechanical leverage is as out-of-place as are anabolic steroids and stimulants on the sports field.

The only valid exceptions are in a few sectors such as statistics, where a mass of routine calculations are necessary in any realistic problem. Yet as these are only part of the syllabus, questions on them should be grouped in a practical paper. In others candidates can show whether they have the ability to handle concepts accurately and fluently, rather than punch out the answers from preset programmes. Even here, careful control and limitation of the models admitted must be maintained. The power and built-in programmes of electronic calculators are growing so rapidly with developments in micro-circuitry that equity between candidates (a most important factor in public examinations that have a great influence on career opportunities) demands that equal facilities should be available to all.

Skill in the use of calculators is, like ability to type or write shorthand, an accomplishment of value in later life. Yet the former contributes as little to the development of numeracy as the latter does to learning English grammar or literature. So it is vital that the strong commercial pressure to boost the market for educational equipment should not be allowed to undermine the basic mathematical education of children to the extent that sweets are ruining their teeth, mechanical transport their physical fitness, the goggle-box their ability to read and write, and canned music their ability to make it themselves, as well as driving them deaf.

The failure to instill sufficient understanding of this universal language of science, and of the characteristics of the extremely powerful but limited tools it provides, so that they are used sensibly, reflects the degree to which some producer co-operatives have been allowed to corner and exploit vital sectors of education. Too many teachers go back into schools without experience of how and why mathematics is used in practical applications. Yet it is in these spheres that most ex-pupils must exercise their mathematical talent in the struggle for survival, or stumble and fall for lack of thorough training in its use.

If the inadequacy of mathematical teaching is to be corrected, more members of professions that make extensive use of mathematics in practice must be produced and got back into the schools to counter-balance the bias caused by the interests and views of academic specialists. Without this feed-back into education, handing on information and experience that must be transmitted in mathematical terms will be strangled.

Greater, almost instinctive sensitivity to the power and limitations of mathematics must be cultivated, and diffused throughout society. Otherwise intelligent, informed, democratic control of scientific developments is impossible.

Unless, until the defects in its educational system are corrected, the secular, politial system must carry as much responsibility for the English disease as the national Church. As in the reform of the latter, a co-operative approach offers the best hope of success.

SPEAK, OR DIE

Chapter 12

The urgency

Mechanical transport

Any society must have a coherent set of principles on which to work. These must be constantly brought up-to-date, and made to bear effectively on its decision making processes. The chaos that failure to satisfy this elementary condition causes in a technologically powered society is well illustrated by the problems associated with the development of the internal combustion engine. The forms of transport that it powers are the hallmark of the 20th century; their production lines and suppliers dominate any modern economy.

The efforts of engineers to provide greater freedom of movement need no apology. Better facilities for easy, wide ranging personal contacts could promote better understanding and relationships between people. Yet the social advantages that this invention can offer have been largely offset, indeed often completely outweighed, by the degree to which human arrogance and greed have been allowed to dominate its application.

There is the dedication of much of the development to increasing the power of armaments. Every master of warfare has recognised the advantages that greater mobility of his forces gives him. So immense efforts have been devoted to utilizing the internal combustion engine to increase the speed, range and accuracy of delivery of the growing destructive power that increasing knowledge of the laws of nature makes available. Without modern transport systems, modern weapons would lose much of their power.

Then many travellers do so for purely selfish purposes. They often impose themselves on the natives or the territory at the other end of the journey, rather than promote mutual understanding. The itch to satisfy these desires has built up pressure to increase the output of vehicles to a self-defeating extent, regardless of the carnage and chaos that intemperate utilization brings in its train.

The fact that the greater convenience that cars, lorries and aeroplanes can provide can only be obtained by placing great mechanical power in the hands of those who drive them is ignored. So is the corollary that such power demands high standards of care and competence from its beneficiaries, if it is to be safely and wisely used.

Genuine concern for the freedom of the individual gets entangled with prejudice, a sub-conscious fear of the magnitude to which problems are allowed to grow before they are noticed, and the choice of inappropriate analogues. So natural rights (if any) get mixed up with artificial ones.

The right to drive or fly anywhere is eagerly claimed. The full cost of building transport systems with all the ancilliary facilities necessary for handling the traffic, without destroying many communities and environments is ignored. The obligation to meet it is rejected.

To stimulate the cancer of 'Growth', styles of promotion stress fashion and snobbery rather than utility and convenience. Designs are changed frequently to provide rapid wastage and jobs for the boys, rather than high efficiency and long life.

As a final act of hypocrisy, the rising toll of life and happiness that must be paid for our failure to face the implications of our wants and natures is classed as accidental. Sometimes even as 'Acts of God'.

Ecological balance

Yet in truth many of our present rights and privileges are not natural, but only tenable as members of disciplined, rational communities. So if mankind cannot manage such a basically simple device as the internal combustion engine sensibly, serious doubts must persist about whether it can handle more complex and powerful processes such as nuclear or biological engineering safely.

The limited, vulnerable biosphere in which we now know that the only hope for man's survival lies is already littered with examples of his refusal to pay the full price of satisfying his desires. These mar the joy of living and threaten the health of growing multitudes. Their rectification involves vastly greater costs and problems than their prevention would have done.

Industrial and domestic wastes still turn many of our rivers and coastal waters into open sewers. Poisonous wastes are spread throughout land, sea and air. Mineral workings and tips, industrial complexes and nuisances are impinging on more and more communities, and steadily reducing greater areas to physical and social deserts. The space necessary for civilised patterns of behaviour and the remaining natural sanctuaries are being eroded.

The derelict and tawdry condition of many urban regions, and the manifest urge to escape from or destroy our man-made environment reveals how mean were the objects on which our natural heritage has been frittered away.

There is massive evidence of the harm, material as well as social, that the measurement of progress by material standards alone causes. So it is surprising that there is not already much deeper, more general and urgent appreciation of the need to re-inject into our thinking and planning a control system commensurate with the scale of the problem.

For these diffuse but extensive indications of the folly of the cult of exponential growth — the Panglossian genuflection to the last and most comprehensive of human vices, covetousness — have been reinforced by more dramatic ones. Yet these have been similarly ignored. The intense savagery of the Western Front is now fading from grandparents' memories. Auschwitz and Stalingrad, Hamburg and Hiroshima are slipping from parents'. The growing callousness with which sectional conflicts are pursued should nevertheless remind men and women how readily technical expertise shorn of moral controls destroys and enslaves.

Yet the urge to go on building even bigger empires is unchecked. The full cost of the biological and social damage done by the first industrial revolution's drastic modification of human living conditions and intensive over-drawing on the Earth's accumulated reserves of energy and raw materials has never been properly evaluated. The problems have been swept out of sight into submerged classes or overseas; and the advent of nuclear power has been harnessed to intensify the process for another generation or two.

More deadly poisons are being produced in increasing quantities. The increase of both population and living standards is being encouraged at accelerating rates, with the same indifference to the welfare of other people and generations as that which betrayed the first attempt to apply science generally for the benefit of humanity into the creation of existing slums and industrial wastelands. The possibility that this may cause some of the problems that already harass the industrial nations to become quite unmanageable is ignored. The probability that, without a higher feed-back and lower rate of growth, the process may become divergent and completely unstable is suppressed.

Feasibility of control

Yet a sticky end is not a necessary consequence of trying to apply scientific techniques to industrial and social problems. If it comes, it will stem from the failure to respect the full implications of the decision. In particular from failure to recognise that such a powerful agent cannot be controlled by attitudes retained from more primitive phases commanding less powerful apparatus.

No single sector of mankind carries all the blame for this failure. The very diverse nature, and differing time scales of classes and nations as they stumbled or were pushed into the secret, and essayed to adopt appropriate styles to exploit it, made prompt recognition of the problem and any combined reaction to solve it exceptionally difficult.

Much of the responsibility for the neglect of the conventions that underpin present prosperity and hopes must be assigned to the indifference or sheer wilfulness of those who have profited most. Yet perplexity also accounts for much. The application of science has already changed, and is still changing, many situations and one's environment so rapidly and profoundly that it is often difficult to see how principles to which one assents in theory should be applied in practice.

Such turbulence is frequently a characteristic of the middle, inflexion stage of a change of state. This is the phase through which the longer industrialised countries are passing. During this period, while some long term trends and constraints can be perceived, it is impossible to clothe them precisely in the numerical garb that is nowadays considered essential for intellectual respectability. Nor are the data adequate for the fine tuning of reactions to current affairs which some pretend that scientific management makes possible.

Yet some factors that affect the coherence and stability of the system can be identified, and the general thrust of their influence detected with sufficient accuracy to allow more intelligent direction of effort.

Acceptability

One drawback to the suggestion that reforming and then

applying more thoroughly, albiet more prudently, a proven system may offer better prospects of coping with a novel situation than any revolutionary prescription is that it sounds rather pedestrian. Yet science often leads to the discovery that things that have evolved quietly and naturally have survived and flourished because they enjoy some intrinsic advantages. Even if the full extent of these, and the reason why they are better than competing developments, is only slowly realised, bit by bit. Re-inventing the wheel (noticing the utility of the wheel in new or unfamiliar circumstances would be a more accurate but less flattering description of the sport) is a constantly recurring theme in the story of how scientific methods have penetrated, demonstrated their effectiveness, and eventually reduced more and more fields of study to order.

Yet however theoretically reassuring such reflections may be, by themselves they offer little firm ground for comfort. The same history also records the reluctance of men to behave, in time and for any length of time, in a sensible and considerate fashion. It reveals a recurring human preference for transient, emotional satisfactions, and the rejection of courses that confer greater benefits, but over a longer period.

Moreover, the extent to which the most popular and powerful means of communication are now locked on to the pursuit of immediate sensation and novelty makes the promotion of rational policies very difficult. The suggestion that a potentially catastrophic situation might best be avoided by the cumulative effect of a series of small, personal changes conflicts with the demand by the media for a constant supply of 'epoch making' revelations for their trumpets.

Such misunderstanding of how the universe works further illustrates how slowly and inadequately the truly epoch making scientific advances of the last century are being assimilated into our culture, even in many schools of science, which should be their main exponents. The cult of the 'hero', whether defined as an individual or a class, and of the 'definitive' break-through, offer such dramatic possibilities to the unscientific artist that they are exploited to the exclusion of more realistic scenarios. So the significance of the work of the real heroes, and the importance of the relationship between them, their companions, and the environment that fostered their activities, has not yet permeated the thinking of many sectors of society. Yet development throughout the whole community of the depth of sensitivity and insight that these can inspire are much more vital contributions to the general welfare that scientific education should make than bemusing the public with facts and formulae.

At times the lemming-like behaviour of people and nations must dishearten any observer. It arouses fear that humanity is so badly infected with the seeds of its own decay that it will be unable to recognise its distress, or muster sufficient courage and energy in time to overcome it. This feeling is reinforced by consideration of the long time constants of processes that must feed back through educational channels. Yet there are contrary indications. If at present they appear no bigger than a hand on the horizon, relief has been known to develop with astonishing speed from quite small beginnings. Within the profession, there is a growing interest in the social responsibilites of science, stretching from *Pugwash* at the highest levels to a slow infiltration of technical assistance for the parish pump. Even the growth of 'consumerism' might reflect a tardy recognition that man was not created simply to meet production targets, inflate statistics, and minister to the prestige of small oligarchies. 'Goods' should be distinguished from 'chattels', and the former epithet restricted to material elements that really do contribute to the welfare of mankind.

Not that such simple, down-to-earth, sensible considerations will have much impact on some groups. Many romantics will still prefer the heady 'Götterdämmerung' attitude of flogging on regardless, just for the fun of seeing what happens in the last few dramatic moments before their bloody but unbowed heads are engulfed in the flames. Trimmers will continue to resent and reject any suggestion that they should stand up and be counted, and fiddle away hoping that something will turn up to save them from having to accept the responsibilities that go with their present level of knowledge and affluence. Yet it is not rational, not in the least scientific, to go on ignoring indications galore of the onset of dangerous conditions, and obstinately refusing to take advantage of the only control system with a known capacity for dealing with such situations, and holding progress within safe limits.

A safe way forward

For if our one, world spanning, scientific civilization is not to generate stresses that could either shatter it to pieces or send it spiralling down into another Dark Age of chaos and oppression, it must have a control system that provides both the necessary indications of aim and performance, and also a servo-power able to restore any divergence between the two.

So this exercise will stick to the only system with a proven track record that offers any rational expectation of a rewarding future for the human race. This is seeing how the Christian attitudes and principles, which enabled men to develop and apply scientific methods thoroughly enough to allow them to change their environment and relationship with it so fundamentally, can be deployed to ensure that they retain control over the civilization they have manufactured, and direct it in a way that does not lead them into slavery or oblivion.

No soft options can provide the clarity of vision and discipline needed for finding the way through unknown territory. There the difference between courage and foolhardiness, caution and pusillanimity, fact and opinion, is always hard to determine, and impossibly so without some well tried instruments at hand, and some reliable, established reference points from which to work.

Chapter 13

The Word

The very seriousness of the present dangers may be the factor that will stimulate sufficient interest to modify disastrous trends. It may be that the immediate and intense pressure of a major crisis is essential for polarising the determination of a large enough proportion of those concerned for effective action to be instigated. One feature of this development must be a dawning realisation that applied science — which has been hailed by many as a panacea — has been diverted from its originally proclaimed aims by attempts to hive off its metaphysical foundations as an irrelevant anachronism. In fact these are needed both to give the system coherence, and prevent its prostitution. In many developed nations, the quest for power, material wealth and self-indulgence are now the sole objectives of large sectors of the population. Engineering has run on under its own momentum, and with enthusiastic encouragement by other interests, to where it is promoting sloth and the manipulation of people, instead of lightening real burdens and cutting slavish chains.

Although there are still many regions where material progress is needed to release men and women from disease and drudgery, in some to survive at all, it seems probable that in the more highly developed countries St. Benedict's prescription of a reasonable quota of hard work for the good of one's soul needs more serious attention. Physical and social limits to providing this for growing populations by producing greater quantities of more sophisticated 'goods' that earn higher wages for buying more 'goods' are already visible on the horizon, even if the ultimate limits of the process cannot yet be accurately determined. Hard thinking and radical reforms are needed to cope with this situation.

False trails

The danger of confusing an invitation to dominate and fill the planet with a licence to overburden it and lay it waste — by encouraging excessive populations and making monetary rewards encashable in material consumption the primary criterion of success — needs correcting before it destroys humanity. The emphasis in human objectives needs shifting from quantity to quality. The political struggle for power and the pursuit of personal affluence lead directly to fear and worry, and the destruction of the environment. What is needed is a balanced pattern of work and recreation conducive to the sound development of individual personalities and personal relationships.

Even by material standards, neglect of the moral basis of the art leads rapidly to bad engineering. It encourages schemes designed to yield short term profits or political kudos at the expense of those that cater for more than immediate demand. These have to pay by a slower return for the greater and more widely distributed advantages they bring, and the reduction in the harm they do.

This will be a difficult reform to carry through, when man is already pressing on the limits of his habitat. Awkward people and problems can no longer be transported to frontier regions. So all are concerned to hold what they have, and are terrified of being left behind in the Gadarene stampede. Yet unless a change of course is executed successfully and fairly promptly, the social stresses that failure to do so are already causing will continue to build to an extent that will endanger the advances that have been won by the application of scientific methods, let alone any hope of further progress.

The main threat to humanity does not stem from those obvious, dramatic, physical manifestations that are most widely, noisily and violently publicised. These are but symptoms of the underlying disease, which is the greed and lust in men's hearts. While treatment of the spots to ease pain and prevent disfigurement is not to be despised, the plague will only be brought under control by attacking the cause directly, and taking steps to eradicate features of the social environment on which it feeds and breeds.

Treatment cannot even start as long as the evidence linking cause and effect is ignored; and everyman contributes to the growing tension by insisting on doing things his own way. It depends on recognition that we are part of a rationally ordered creation. Then restoring authority to the Spirit that orders it, and doing things his way.

The right way

If Christian ideals are to render this service to mankind, they must be imprinted boldly in codes of practice and decisions. This cannot be done by vague sympathies and fading memories; there is too much deliberate selfishness to be contained.

For although the creed includes within its ambit the simplest domestic virtues, Christianity is not just a way of teaching the children to be kind to Grannie and the cat. It is a way of life that demands responsible, dedicated commitment limited only by each individual's own endowment of abilities. It challenges the strong as surely as it comforts the weak. Although the lily and the dove occupy honoured places in the Church's array of symbols, it is the Cross that must always dominate its witness. It demonstrates to mankind an unbounded love that calls for a response, and reveals a light that can redeem even the foulest human folly.

Christianity is neither a set of social norms to which people must conform for their own safety and to please their creator, nor formulae for rationalising the relationships that appear or develop between a mass of purely physical units and assemblies. It asserts that personality is the real, recognisable if indefinable feature that gives men and women a unique role in the universe, and that the purpose of their creation is to enjoy a pattern of personal relationships with their Creator and their fellows described as loving, in the unadulterated sense of the epithet. It is this element in them that justifies the belief that their lives have some lasting value.

Yet this can only be achieved if they will learn the lesson of the Cross — that sacrifice rather than self-satisfaction is the heart of the matter — and accept its offer of reconciliation with the only power that can enable them to overcome their inherent tendency to forget it, and blunder into trouble.

This belief does not in the least disparage human achievement in the physical and social sciences. The solution of problems in these fields is for many, and in many situations, the most effective, practical method of expressing love, by helping over-burdened neighbours here on Earth. But it does demand recognition that the attainment of any level of scientific knowledge, material prosperity or political development is only a very poor, secondary indicator of welfare or progress. Like all indirect measurements, these can be most misleading unless they are used with the greatest discretion, and constantly recalibrated by reference to more fundamental criteria.

Jesus was not only the way by which the European community was led to the discovery of modern science. He is the only Lord who has, and has demonstrated, an authority enlightened and comprehensive enough to enable a super-critical assembly of human beings to harness the Great Sources of Power in Nature successfully, and direct them roughly towards a dimly perceived vision of the true welfare of mankind. And the guide who led the communities that initiated this movement to and through its take-off phase is even more necessary during the very delicate manoeuvres that are now required if Earth's economy is to be re-established at whatever higher level applied science can sustain it, without in the meantime destroying the ecology and environment essential for its survival. Technical processes must respect the very purpose of existence if their beneficiaries hope to survive at all.

Having unleashed the intoxicating flood of scientifically generated development, technological man is now in danger — like the Sorcerers's Apprentice — of being overwhelmed by a process he cannot himself control. He cannot put off much longer coming to terms with the reality of his predicament, by ever more extravagant waste of irreplaceable assets, and by trying to sweep problems out of sight and mind into backyard areas. The supply of both is running short.

He must speak, or die. Yet before speaking, he is manifestly well advised to equip himself with the right Word.

The Word

This is not difficult to find. One of the genuine benefits that has flowed from the application of scientific methods to communication and production problems is that the Bible is now open and intelligible to more people than ever before. So few are barred from finding the Way to more abundant life, if they want to do so.

Man cannot escape from the truth of the well proven, scientific as well as moral law, that responsibility goes hand-in-hand with power and influence. As the latter grow, so does the former. Having gorged himself so fully on the fruit of the tree of knowledge, he cannot pass the buck for choosing between good and evil, blessing and curse, life and death to anyone else. It sticks quite firmly in his grasping hands. It would seem sensible to choose the first alternatives, rather than slip into the second, whether deliberately or by default.

Yet the nature of the creation he enjoys has not only determined the questions that he must answer, and the options available. His guided growth to a dominant role that can rightly be described as potential 'children of God' means that the choice is his. To say 'yes' to the offer of reconciliation with this relationship, and live; or turn away, and die.

Chapter 14

At present

'We shape our buildings, thereafter they shape us.' — Winston Churchill.

Engineering has played a major part in building the environment that is now moulding our lives, and the direction in which further progress is pointed.

The structure of the modern city is largely dictated by the means of transporting men and materials that were available at its conception. The commerical city is dominated by office concentrations, and communication systems for gathering people and information. The industrial city reflects the needs of production lines. The style and intensity of each depends on the building techniques available, and the service facilities they need to operate.

The countryside has been drastically remodelled by mineral extraction, and agricultural systems keyed to mechanical handling and chemical processing. It is festooned with growing service and communication networks.

The technical justifications for enormous, costly and unmanageable urban concentrations are now obsolescent. So it is not surprising that many are becoming festering pockets of decay and unrest. They need reducing to balanced, manageable proportions.

While some reduction in the manpower demands of agriculture and increase in its yields was once beneficial, the trend towards mechanised ranching has gone too far, socially as well as ecologically. It has isolated human beings too completely from their natural heritage and background.

Policies that sustain the maximum number of small, almost minimal units in both industry and agriculture are needed.

Yet the influence of social structures is even stronger than that of physical ones. Indeed it is often pressure from social groups that encourages the exploitation of technical expertise beyond mankind's ability to control the consequences. Which reflects badly on advisers who claim to be skilled in system analysis, design and control.

Engineers must respect the authority that gave their profession its strength and opportunity, by restoring its moral purpose into their operations. Otherwise others may justly accuse them of being the barbarians responsible for most current woes. The correct starting point must be their own structures and relationships.

This leads to asking questions, rather than proffering ready-made answers. For during a transition between states, adequate data on which long-term solutions can be reliably based is not available, whereas recognition of some of the issues is feasible. So drawing attention to how these should be approached may be the best contribution the profession can make to the welfare and survival of mankind during the present, very turbulent, psychosocial phase of its evolution.

In the profession

There is no lack of interest in raising the quality and status of engineering in Britain today. Proposals for achieving this usually take the form of tighter, more formal control of entry. These lead to closed shops, similar to those enjoyed by the ancient professions. Is there any evidence that this would improve the service that engineering offers society? That the service offered by the preferred exemplars is better? Even adequate? That the service given by British engineers in past has been worse than that given by more formally organised systems? To deny the claim that the existing, less rigid or hierarchical pattern is an advantage?

The performance of British engineers, despite their well attested low social status, supports the latter contention. Many of these entered the profession, achieved great things, enabled the nation and many of its industries to survive, despite political incompetence and poor rewards, by paths that would be closed by the restrictions now proposed. How would barring such men improve the position? The suggestion that our present, or any, educational system can identify them early and with certainty is nonsense. Might the proposals be just one more example of the fashionable trend to grant dangerous monopoly powers to producer co-operatives? And of bureaucratic indifference, even hostility, to the true interests of *all* the people?

Might engineers be being tempted to repeat the errors that have hamstrung the ability of the ancient professions to provide the services expected of them, with which they are charged? Their associations seem more concerned with their own interests than with those of the community. Might the failure of such groups, which should provide sound advice and leadership to the community, be a major cause of Britain's descent into anarcho-syndicalism? So is it a precedent that scientific professions are entitled to follow? Particularly those commited to serving the use and convenience of their fellows by the application of rational, scientific methods?

In service

Does a similar trend towards the creation of self-serving rather than socially sensitive structures afflict the engineering based industries. Are projects and plans really selected for the use and convenience of mankind, or to inflate the egos and incomes of insiders? Two aspects seem very questionable. A high rate of change obviously provides high levels of employment and reward for those who devise, produce and sell new products. Is the rate of change now being imposed tolerable by any community?

Might the widespread social unrest in the advanced nations, often verging on anarchy and chaos, be evidence of trying to maintain a higher rate of change in the environment and human relationships than any system can assimilate without breaking down? And is not much of the novelty that sustains the process trivial, a mere waste of effort and materials to give the illusion of progress?

Many developments require massive institutional arrangements. Are these becoming, like the brontosauri, too big for sensitive control and prompt, flexible reaction to changing circumstances? Conditioned for success against their peers, but more vulnerable than less cumbersome species to a change in climate? Often so big that they slip, or are pushed, into monopolies?

When most of the information about and opportunities for service in any technical field are concentrated within a single hierarchy, has the problem of maintaining the integrity of advice in, and hence democratic control over, that sector been adequately considered? And solved?

In training

Might the failure of engineers to show signs of recognising and tackling obstacles to directing their art for the use and convenience of mankind be a significant factor in the present tendency of many bright, altruistic young people who would do well in the profession to shy away from the study of the discipline? Hence crippling efforts to improve the entry to it? And incapacitating the rising generation from comprehending and influencing events that, in a fuzzy way, it feels are driving mankind in a dangerously wrong direction? Without deep, thorough understanding of the creature that is carrying the human race somewhere so fast, the rider has little chance of controlling it. Wishful thinking will not stop a runaway truck, some knowledge of mechanics might.

The most serious aspect of the University Grants Committee's discrimination against the leading technological universities is not its incompatibility with other governmental plans for reviving the economy (they are flawed by the doctrine of growth) but its toleration of the bias towards literary studies and soft options. These isolate too many from contact with scientific methods, and then gaining some useful, practical experience in industry and commerce. Many more with this background must be fed back into fields such as schools, homes, churches and political parties, to enable society as a whole to understand the system on which it is relying for its progress and survival.

Can this barrier to effective internal communications, and the danger of social fission it causes, be obviated unless the university schools of science (pure and applied) accept, indeed insist on, a genuinely more liberal, less specialised school curriculum? Instead of using their control of entry to force candidates to devote the last 2/3 years of schooling to the preliminary part of degree courses? Reform would manifestly lengthen higher education. Is submission to administrative pressure for minimun first cost leading to shoddy education as surely as it did to jerry-built housing?

But do educationalists spend what time and money is provided wisely? Are they encouraged to do so by practising engineers? Or are the latter looking for a subsidy? For schools to be forced to try to do professional training (and do it badly) that is the responsibility of industry? One can only learn a job by doing it.

In society

Can engineers expect their professional activities to be fruitfully and safely exploited if they themselves do not take a full, active part in the systems of government and guidance? Can an administrative system that excludes engineers from its policy forming ranks hope to control a technologically based economy successfully? Can a Parliament with an inadequate engineering component legislate for and control the administration of a society dependent on such an economy effectively?

Can Churches with hierarchies devoid of engineering experience hope to offer any useful advice on how the creeds they recite apply to highly developed, scientifically sustained societies? Was Jesus just wasting his time for most of his adult life?

Scientifically, it is nonsense to imagine that such a formidable set of problems can be adequately solved without the engineers concerned paying just as much attention to the maintenance of the quality of the moral standards used in measuring the personal and social aspects of their performance as they do to the mechanical ones used in the physical aspects. Unless they do accept their due share of responsibility for this work, it will be botched. Which would be unfortunate, as it has a strong bearing on the welfare and survival of both the profession and the community it serves.

Chapter 15

Growth?

The popular fantasy of 'Growth' — the illusion that some political gimmick can achieve the exponential growth in material wealth reflected in the demand for annual increases of x% in communal output and (x+n)% in personal rewards — has become the 'cargo cult' of the developed nations. Engineers have played a major part in stimulating this latest attempt to bring heaven physically down to earth. Belief in its feasibility owes much to misinterpretation of mathematical data. As pursuit of this chimera poses a serious threat to consolidation of the real progress that has been made, and even human survival, an indication of its folly seems an appropriate epilogue to this enquiry.

Confirmation that the unlimited growth implied by this doctrine is unattainable is one of the most valuable, and least remarked, bits of hard intelligence already won by man's latest technical triumph, space travel. Its practical realisation has given an indication of the order of magnitude of the cost of transporting men and materials between planetary regions. Returning vehicles have given an indication of the economic potential of those within reach. This information is similar to that on inter-continental relationships won for Europe by Columbus and other early explorers. Comparing the infinitessimal economic value of the cargoes brought back and the sterility of the areas visited with the cost of space travel, and contrasting this with the immediate return from vast, fertile, rich territories discovered for a very modest outlay formerly, shows that this great achievement is — like so much military expenditure — an economic sink.

For all normal purposes mankind inhabits a strictly limited territory. All the factual evidence is against any El Dorado lying within exploitable range that might offer the whole Earth space and resources for expansion comparable to those that the opening up of Africa, America and Australasia has given Europe during the last few centuries. Science fiction's vision of satellite Utopias are as reliable a guide to the characteristics of space as 'Alice through the Looking-Glass'. Comparing the cost of reaching, establishing and sustaining such colonies with the undeveloped state of much more accessible and hospitable regions like Antarctica and the Sahara demonstrates how unrealistic such projections are.

Within a finite containment, the infinite growth implied by an exponential function is manifestly impossible. Malthus had the right idea. Even if, lacking information since acquired and the more powerful machinery now available for handling it and identifying critical paths and components, he named the limiting factor wrongly. This is very easily done when pioneering, with inevitably scanty knowledge of many of the parameters involved. It now looks as if this may be habitable space itself — for accommodating the scale of food, energy and material production, processing and distribution required by the standard of living expected nowadays. And to enable people to live in decency, without getting on one another's nerves, at one another's throats.

The threat has long been disguised by the extent to which European nations — the first, and until recently the only exponents of scientifically based growth — have been able to expand into lightly occupied areas in other continents for living room. They have exploited their resources to maintain the process after they had overloaded or exhausted their indigenous resources. Yet as other nations follow their example, the ratio of reserves to expectations is falling fast.

Most economists and politicians have confused two functions that have similar traces over part of their range. One can fit much economic data over several generations to logarithmic against time charts. This might indicate an exponential relationship; but it might equally well record the first half of a change of state. In an autonomous system, the initial rate of change during a transition is often proportional to the change that has already taken place. This is the driving force. So at this stage the track followed is exponential. The constraints that will eventually modify it may not be generated or brought into action until sufficient growth has occurred for their need to be apparent. Only then does their influence become detectable.

As mankind is but a dozen generations (of the dominant species, the correct unit of time in analysing a biologically controlled process) from the start of the movement that has caused the current change of state — from a society dependent on traditional methods of production, to one where development is speeded and intensified by the application of scientific methods — one would not expect the divergence between the two traces to be very obvious yet. Particularly in interpreting the widely scattered information that must be integrated in measuring the total capacity and tolerable standard of living of a whole planet. So other means of ascertaining whether limits to growth exist are needed.

Hence the importance of the evidence from space. Moreover, this is confirmed by the instability that afflicts the economies of the advanced nations, social conditions in the conurbations that are characteristic of them, and the international relationships on which they depend. These are typical signs of a system moving into the later stages of a change of state so out of control that it is likely to plunge to destruction or veer off into oblivion before it can be re-stabilized at whatever higher level would otherwise have been sustainable.

Closely allied to this consideration is whether there are limits to the rate of change that any human community can tolerate without disintegrating under internal stresses. Like currents and rocks in strange, uncharted waters, accurate knowledge of the nature and distance of such hazards is not available before they are reached. Yet once there is reason to suspect that they exist, it is sensible to approach situations that indicate the possibility of their presence with some caution.

Populations have already been allowed, in many countries and by many pundits encouraged, to grow to such an extent that the planet cannot maintain the vast majority in the state that they covet as they watch the more affluent sectors on their screens. Even if some further growth is desirable and maintainable in the average standard of living in the long run, this will have to be mainly in intangible assets, such as knowledge, artistic expression and reflective leisure. The utilization of land for all purposes, and the consumption of irreplaceable material resources — figures for which are commonly used as indices of economic achievement — will have to be at a slower, decelerating rate, tending towards zero and a balanced system of re-cycling. Major changes in people's attitudes will be needed if such a state is to be reached, and the fatal drive of lust and greed controlled. The scale of educational effort that will be required is daunting. Especially when so many aspects of the situation must be encoded in mathematical language, which at present so few can comprehend.

Despite the advances since Malthus' time that have modified his predictions, struggles for resources have already caused massive cullings of people by war, famine and disease. The development of even more powerful methods of destruction continues apace. So radical reforms are now urgently needed, if the human race is to use the high degree of influence over its pattern of evolution that its highly developed level of mental and social organisation bestows on it in ways that are compatible with its own welfare and survival.

The task will not be accomplished unless the futility of raising mechanical structures on unsound moral foundations is recognised.